Crop Production Science in Ho...
series

Serie
Unive
Cons

This se
the ma
areas.
tected
placed
than
unders
solutio
St
in cou
and ap
as a s
valuab
author
garder
will als
Th
experie
aspect
and pla
storag
inform

Titles
1. **Orn**
2. **Citr**
3. **Oni**
4. **Orn**
5. **Bar**
6. **Cuc**
7. **Tro**
8. **Cot**
9. **Let**

Carrots and Related Vegetable *Umbelliferae*

V.E. Rubatzky,
C.F. Quiros
Department of Vegetable Crops
University of California, Davis
USA

and

P.W. Simon
USDA-ARS
University of Wisconsin, Madison
USA

CABI *Publishing*

CABI *Publishing* is a division of CAB *International*

CABI Publishing
CAB International
Wallingford
Oxon OX10 8DE
UK

Tel: +44 (0)1491 832111
Fax: +44 (0) 1491 833508
Email: cabi@cabi.org

CABI Publishing
10 E 40th Street
Suite 3203
New York, NY 10016
USA

Tel: +1 212 481 7018
Fax: +1 212 686 7993
Email: cabi-nao@cabi.org

A catalogue record for this book is available from the British Library, London, UK.

Library of Congress Cataloging-in-Publication Data
Rubatzky, Vincent E.
Carrots and related vegetable Umbelliferae / V.E. Rubatzky, C.F. Quires, and P.W. Simon.
p. cm. -- (Crop production science in horticulture series; 10)
Includes bibliographical references (p.)
ISBN 0-85199-129-7 (alk. paper)
1. Umbelliferae. 2. Vegetables. I. Quiros, C.F. (Carlos F.). II. Simon, P.W. (Philipp W.) III. Title. IV. Series: Crop production science in horticulture; 10.
SB351.U54R83 1999 99-26900
635′.13--dc21 CIP

ISBN 0 85199 129 7

Typeset in Britain by Solidus (Bristol) Ltd
Printed and bound in the UK by the University Press, Cambridge

Contents

PREFACE

This book was written with the intention to bring together background information relating to the history, descriptions, production and uses of carrots, celery and other umbelliferous vegetables. Although not a detailed production manual for each of these diverse crops, many principles and insights are included that can assist production practices, particularly for carrots and celery where the genetic and environmental factors influencing crop performance and procedures for optimum handling are fairly well developed. We have indicated the influence of environmental factors on crop production, and given consideration to plant characteristics that affect performance ranging from seed germination to reproduction. References cited can lead those wishing for more detail to other sources of information and research. Beyond carrots and celery, we also wanted to present the interesting diversity of these crops and their usage, many of which are very limited in production.

Umbellifers have received relatively little research attention compared with other vegetable species such as tomato, the brassicas, cucurbits and alliums. The umbelliferous vegetables are an interesting group of plants generally characterized by strong and unique flavour compounds, and in some cases providing important nutrients which can enrich the diet of consumers. For example, the nutritional value of the carrot is understated and unfortunately frequently is an under-utilized source of pro-vitamin A, especially in developing parts of the world in subtropical and tropical regions. Greater consumption of carrots and other umbellifer sources of carotenoids would measurably improve the health of many people who are presently deficient in vitamin A, as well as enhance the enjoyment of their meals.

In general, our discussion centres on carrots or celery and one can usually assume that the information is relevant to other vegetable umbellifers. The reader will find some areas discussed in more detail than others due in part to the individual interests and expertise of the authors. We believe that being able to feed off each other's experiences was beneficial. The absence of information on certain potentially useful areas also became apparent. One of our discoveries during preparing this book was that there was a conspicuous lack of information available about many of the umbelliferous vegetables, especially regarding crop production. There was also a lack of information about fundamental biological attributes such as anatomy, physiology or genetics. New technologies have the potential to greatly broaden our knowledge about these vegetables and open new areas of information. It is logical to expect considerable progress as new technologies are employed to develop further information about and the utilization of these vegetables.

Although our mission was to focus on the vegetable *Umbelliferae*, the distinction between vegetable and spice brought us to a discussion of crops like coriander where the leaves are used as a vegetable and the seeds are used as a spice. We hope that our inclusion of these dual-use crops does not dissuade other authors from covering them.

We have visited many references, those indicated and numerous others that also provided support. We especially want to acknowledge researchers at HRI and the earlier NVRS in Wellesbourne who were instrumental in greatly illuminating many aspects of carrot and celery biology.

The authors are grateful to CAB *Publishing*'s Editorial Assistant Ali West whose extraordinary patience with this long delayed effort enabled its completion. Appreciation is certainly due to our wives Verna Rubatzky, Sandy Larson and Raquel Quiros who permitted us the 'extra time' that should have been theirs, in order that we could participate in this effort.

1

Introduction: Geographic Origins and World Importance

INTRODUCTION

The botanical family *Apiaceae* (previously known as *Umbelliferae*) consists of about 250 genera and approximately 2800 species of widely distributed, generally herbaceous plants, mostly growing in temperate and boreal regions. Adoption of this name change will take time and many readers will continue to recognize *Umbelliferae* rather than *Apiaceae*. In that context, we use 'umbellifer' to refer to plants in this family throughout this book.

Among the vegetable umbellifers are plants with annual, biennial and perennial growth characteristics. The most distinctive characteristic of the family is the inflorescence, properly known as an umbel, derived from the Latin *umbella*, meaning sunshade. The typical umbel is a convex or flat-topped flower cluster in which individual pedicels all arise from the same apex. Other important characteristics of vegetable umbellifers are their fruit, a schizocarp containing two mericarps, and unique biochemistries that are responsible for the distinctive flavour and aroma of various plant parts.

Considerable diversity exists between *Umbelliferae* species. For example, *Azorella* is a genus containing compact fibrous plants that have in the past served as a primary fuel source for the highland natives in the Andean mountains (Hodge, 1960), whereas poison hemlock *Conium maculatum* is a rare alkaloid-containing plant, well known for the poisoning of Socrates. Substantial diversity is seen among vegetable umbellifers grown for their edible roots, tubers, stems, petioles, leaves, flowers, fruits, seeds, extractable essential oils, carotenoids and other compounds. Several crops such as carrot, celery, parsnip, cilantro and arracacha are considered among the major vegetables in certain regions. Many others are produced on a smaller scale, or mainly in home gardens, and some are not cultivated but gathered from the wild. Relatively few vegetable umbellifers are high caloric contributors, but they often excel in providing micro-nutrients and a broad medley of flavouring and textural characteristics directly as foods and as additives that enhance the enjoyment of other foods. We have included several umbellifers that are used primarily for flavouring, condiment and/or garnish because they have various vegetable-associated purposes. Many cultivated and wild umbellifers have a long and extensive history as folk medicines, although generally not used in contemporary medicine.

A listing of edible vegetable *Umbelliferae* by common and botanical names, their primary uses, and plant portions utilized appears in Tables 1.1 and 1.2.

CROP ORIGIN AND DOMESTICATION

The origin and domestication of many vegetable *Umbelliferae* crops are frequently associated with the Mediterranean and nearby Asian regions. Some exceptions are: arracacha and culantro (Central and South America); Japanese hornwort (China, Japan and South-east Asia); water dropwort and lawn pennywort (South-east Asia). Samphire and sea holly originated in coastal areas of northern Europe. American angelica and epos have North American origins.

Food and/or condiment use is made of different plant parts. These include root, tuber, stem and leaf tissues, and seed products. For many plants several plant parts, such as leaves and roots can be consumed, and for some, essentially all plant parts can be utilized as a food or condiment.

Root crops

Several vegetable umbellifers are grown specifically for their edible storage root and tuber portions, carrots being the most important. In addition to carrot, other umbellifer root crops include: parsnip, celeriac, arracacha, Hamburg parsley, turnip-rooted chervil, skirret, asafetida, great earthnut, cow parsnip, greater burnet saxifrage, epos and caraway. Some of these also have a secondary use for their edible foliage and/or seeds for condiment usage.

Carrot (*Daucus carota* L., var. *sativus* Hoffm.) is the major vegetable umbellifer cultivated worldwide. Cultivated carrots can be separated into two types: eastern/Asiatic and western. Eastern/Asiatic carrots have reddish purple (anthocyanin-containing) or yellow roots, pubescent leaves which give a grey-green appearance, and a tendency for early flowering. Western carrots have orange, yellow, red or white roots, less pubescent green leaves, and less tendency to bolt without extended exposure to low temperatures.

Also widely distributed is the related weedy species *D. carota* var. *carota*. Vavilov (1951) reported wild carrots to be abundant in Afghanistan and Turkestan (Central Asia) as well as distributed from the Atlantic coast of eastern Europe to western China. He regarded the origin of the Asian cultivated carrots to be in the Inner Asiatic Centre, and the origin of western

Table 1.1. Edible vegetable *Umbelliferae* species, listed by common and botanical names, their primary uses and plant portions utilized.

Common name	Botanical name	Uses*	Plant portion used[†]
Ajowan	*Trachyspermum ammi*	C	S, L
Ajwain	*Trachyspermum ammi*	C	S, L
Alexanders	*Smyrnium olusatrum*	V, C	L, Stk
American angelica	*Angelica atropupurea*	V	St, S, R
Anise	*Pimpinella anisum*	C	S, L
Arracacha	*Arracacia xanthorrhiza*	V	R, L, St
Asafetida	*Ferula assa-foetida*	C	R, Sh
Asiatic pennywort	*Centella asiatica*	V	L
Bishop's elder	*Aegopodium podagraria*	V	L, St
Bishop's weed	*Aegopodium podagraria*	V	L, St
Bitter fennel	*Foeniculum vulgare* var. *vulgare*	C	S
Black caraway	*Carum nigrum*	C	S
Black cumin	*Bunium persicum*	C	S
Black lovage	*Smyrnium olusatrum*	V, C	L, Stk
Burnet saxifrage	*Pimpinella saxifrage*	C	S, L, R
Caraway	*Carum carvi*	C	S, R, L
Carrot	*Daucus carota* var. *sativus*	V	R, L
Celeriac	*Apium graveolens* var. *rapaceum*	V	R
Celery	*Apium graveolens* var. *dulce*	V, C	L, S
Chervil	*Anthriscus cerefolium*	V, C	L, S
Chinese parsley	*Coriandrum sativum*	V	L
Cilantro	*Coriandrum sativum*	V	L
Common cow parsnip	*Heracleum sphondylium*	V	Sh, L
Common giant fennel	*Ferula communis*	C, V	S, L
Coriander	*Coriandrum sativum*	V, C	S, L
Cow parsley	*Anthriscus sylvestris*	V	L
Cow parsnip	*Heracleum lanatum*	V, C	R, L, S
Culantro	*Eryngium foetidum*	V	L
Cumin	*Cuminum cyminum*	C	S
Devil's dung	*Ferula assa-foetida*	C	R, Sh
Dill	*Anethum graveolens*	C, V	S, L
Epos	*Perideridia gairdneri*	V	R
European angelica	*Angelica archangelica*	C, V	St, S, R, L
Fennel	*Foeniculum vulgare*	C, V	S, L
Finocchio	*Foeniculum vulgare* var. *azoricum*	V, C	L, S
Florence fennel	*Foeniculum vulgare* var. *azoricum*	V,C	L, S
French parsley	*Anthriscus cerefolium*	V, C	L, S
Garden angelica	*Angelica archangelica*	C, V	St, S, R, L
Garden chervil	*Anthriscus cerefolium*	V, C	L, S
Garden lovage	*Levisticum officinale*	C, V	L, St, S
Garden myrrh	*Myrrhis odourata*	C	L, S, R
Giant fennel	*Ferula assa-foetida*	C	R, Sh, S
Great earthnut	*Bunium bulbocastanum*	V, C	T, L, F

Continued

Table 1.1. *Continued*

Common name	Botanical name	Uses*	Plant portion used†
Greater burnet saxifrage	*Pimpinella major*	C	R, S, L
Hamburg parsley	*Petroselinum crispum* var. *tuberosum*	V, C	R, L
Horse parsley	*Smyrnium olusatrum*	V, C	L, Stk
Indian dill	*Anethum sowa*	C	S, L
Indian pennywort	*Centella asiatica*	V	L
Japanese angelica	*Angelica edulis*	V	Stk, L
Japanese hornwort	*Cryptotaenia japonica*	V	L, St, R
Japanese parsley	*Cryptotaenia japonica*	V	L, St, R
Java coriander	*Eryngium foetidum*	V	L
Knob celery	*Apium graveolens* var. *rapaceum*	V	R
Lawn pennywort	*Hydrocotyle sibthorpioides*	V	L, St
Lovage	*Levisticum officinale*	C, V	L, S, R
Mexican parsley	*Coriandrum sativum*	V	L
Mitsuba	*Cryptotaenia japonica*	V	L, St, R
Myrrh	*Myrrhis odourata*	C	L, S, R
Parsley	*Petroselinum crispum*	V, C	L
Parsnip	*Pastinaca sativa*	V	R, L
Purple angelica	*Angelica atropupurea*	V	St, S, R
Rock samphire	*Crithmum maritimum*	V	L
Salad chervil	*Anthriscus cerefolium*	V, C	L, S
Samphire	*Crithmum maritimum*	V	L
Scotch lovage	*Ligusticum scoticum*	V	L, Sh, R
Sea fennel	*Crithmum maritimum*	V	L
Sea holly	*Eryngium maritimum*	V	R, Sh, L
Skirret	*Sium sisarum*	V	R
Smallage	*Apium graveolens* var. *secalinum*	V, C	L, S
Squawroot	*Perideridia gairdneri*	V	R, S
Stinking gum	*Ferula assa-foetida*	C	R, Sh
Stinkweed	*Eryngium foetidum*	V	L
Sweet anise	*Foeniculum vulgare* var. *dulce*	C	S, L
Sweet chervil	*Myrrhis odourata*	C	L, S, R
Sweet cicely	*Myrrhis odourata*	C	L, S, R
Sweet fennel	*Foeniculum vulgare* var. *dulce*	C	S, L
Turnip-rooted chervil	*Chaerophyllum bulbosum*	V	R, L
Turnip-rooted parsley	*Petroselinum crispum* var. *tuberosum*	V	R, L
Water celery	*Oenanthe javanica*	V	L, Sh
Water dropwort	*Oenanthe javanica*	V	L, Sh
White chervil	*Cryptotaenia canadensis*	V	L, St, R, F
Wild chervil	*Cryptotaenia canadensis*	V	L, St, R, F
Wild parsnip	*Angelica sylvestris*	V, C	L, St, R, S
Yampah	*Perideridia gairdneri*	V	R, S

* V, vegetable; C, condiment or flavouring.

† F, flowers; L, leaves; R, roots; S, seeds; Sh, shoots; St, stems; Stk, stalks; T, tubers.

Table 1.2. Edible vegetable *Umbelliferae* species listed by botanical and common names, their primary uses and plant portion utilized.

Botanical name	Common name	Uses*	Plant portion used†
Aegopodium podagraria	Bishop's weed	V	L, St
Anethum graveolens	Dill	C, V	S, L
Anethum sowa	Indian dill	C, V	S, L
Angelica archangelica	Garden or European angelica	C, V	St, S, R, L
Angelica atropupurea	American or purple angelica	V	St, S, R
Angelica edulis	Japanese angelica	V	Stk, L
Anthriscus cerefolium	Salad chervil, French parsley	V, C	L, S
Anthriscus sylvestris	Cow parsley	V	L
Apium graveolens var. *dulce*	Celery	V, C	L, S
Apium graveolens var. *rapaceum*	Celeriac	V	R
Apium graveolens var. *secalinum*	Smallage	V, C	L, S
Arracacia xanthorrhiza	Arracacha	V	R, L, St
Bunium bulbocastanum	Great earthnut	V, C	T, L, F
Bunium persicum	Black cumin	C	S
Carum carvi	Caraway	C	S, R, L
Carum nigrum	Black caraway	C	S
Centella asiatica	Asiatic or Indian pennywort	V	L
Chaerophyllum bulbosum	Turnip-rooted chervil	V	R, L
Coriandrum sativum	Coriander, cilantro, Chinese or Mexican parsley	V, C	S, L
Crithmum maritimum	Samphire or rock samphire	V	L
Cryptotaenia canadensis	Hornwort, white or wild chervil	V	L, St, R, F
Cryptotaenia japonica	Japanese hornwort or Mitsuba	V	L, St, R
Cuminum cyminum	Cumin	C	S
Daucus carota var. *sativus*	Carrot	V	R, L
Eryngium foetidum	Culantro, Java coriander, Stinkweed	V	L
Eryngium maritimum	Sea holly	V	R, Sh, L
Ferula assa-foetida	Asafetida or giant fennel	C	R, Sh, S
Ferula communis	Common giant fennel	C, V	S, L
Foeniculum vulgare	Fennel	C, V	S, L
Foeniculum vulgare var. *azoricum*	Florence fennel	V, C	L, S
Foeniculum vulgare var. *dulce*	Sweet anise	C	S, L
Heracleum lanatum	Cow parsnip or masterwort	V, C	R, L, S

Continued

Table 1.2. *Continued*

Botanical name	Common name	Uses*	Plant portion used†
Heracleum sphondylium	Common cow parsnip	V	Sh, L
Hydrocotyle sibthorpioides	Lawn pennywort	V	L, St
Levisticum officinale	Lovage, garden lovage	C, V	L, S
Ligusticum scoticum	Scotch lovage	V	L
Myrrhis odourata	Sweet cicely, garden myrrh, Fern-leafed or sweet chervil	C	L, S, R
Oenanthe javanica	Water dropwort or water celery	V	L, Sh
Pastinaca sativa	Parsnip	V	R, L
Perideridia gairdneri	Epos, squawroot or yampah	V	R
Petroselinum crispum	Parsley	V, C	L
Petroselinum crispum var. *tuberosum*	Hamburg or turnip-rooted parsley	V, C	R, L
Pimpinella anisum	Anise	C	S, L,
Pimpinella major	Greater burnet saxifrage	C	R, L, S
Pimpinella saxifrage	Burnet saxifrage	C	S, L, R
Sium sisarum	Skirret	V	R
Smyrnium olusatrum	Black lovage, horse parsley or Alexanders	V, C	L, Stk
Trachyspermum ammi	Ajowan or ajwain	C	S, L

*V, vegetable; C, condiment or flavouring.
† F, flowers; L, leaves; R, roots; S, seeds; Sh, shoots; St, stems; Stk, stalks; T, tubers.

cultivated carrots in the Asia Minor Centre, primarily Turkey. Mackevic (1929) and Heywood (1983) suggested the Himalayan Hindu Kush region of Afghanistan as the primary centre of diversity for eastern carrots.

The origin of western carrots has been studied extensively, but little evidence for carrot cultivation exists before the 10th century. Carrot seeds were found with other vegetable seeds in Switzerland and southern Germany dating from 2000–3000 BC. The seed are thought to have been the plant part used, rather than the root. Some historians believe the ancient Greeks and Romans cultivated carrots. However, most ancient writings from Asia Minor, Greece and Rome do not mention carrots, even though wild carrots have a long history of presumed medicinal uses. Interestingly, parsnips were often noted in such writings.

According to other written documents, purple or red and yellow carrots were cultivated in Iran and northern Arabia in the 10th century, and in Syria about the 11th century. By the 12th century, carrots were reported in Spain, followed by Italy (13th century), France, Germany, and The Netherlands (14th century) and England (15th century) (Banga, 1963; Fig. 1.1).

Carrots were first reported in China late in the 13th century, about the time they became popular in Europe. Carrots came to the USA with European

Time	Location	Colour
Pre-900s	Afghanistan and vicinity	Purple and yellow
900s	Iran and northern Arabia	
1000s	Syria and North Africa	
1100s	Spain	
1200-1300	Italy and China	
1300s	France, Germany and The Netherlands	
1400s	England	
1600s	Japan	
1600s	Northern Europe and North America	Orange and white
1700s	Japan	Orange
1721	Northern Europe	'Long Orange' and 'Horn' types described

Fig. 1.1. Origins of cultivated carrots.

settlers in the 17th century, and were first reported in Japan in the 18th century. As in Europe, the first Chinese carrots were purple or yellow rooted. Since leaf appearances were not reported, it is not possible to know whether the first European carrots were comparable to Eastern carrots or, as is more likely, that they had less pubescent leaves. It can be concluded that European carrots never had or lost the early bolting tendency of eastern carrots since production of palatable roots would not have been possible.

Yellow carrots described in Iran and Arabia had culinary quality inferior to purple-rooted carrots. Nevertheless, they eventually replaced purple carrots in Europe. In England, the purple carrot had only a short popularity before replacement by yellow types. Perhaps adaptation for more northern latitudes also brought along improved flavour.

The origin of western yellow and purple carrots is well accepted, but the development of western orange types is less clear. Banga (1963) proposed, on the basis of the appearance of carrots in European oil paintings, that orange-rooted carrots originated in Holland in the early 17th century as selections from yellow carrots. Written documentation of orange carrots did not appear until 1721, with the description of the 'Long Orange' and several 'Horn' types. The first definite German orange-rooted carrot described in 1740 was 'Brunswick', although Banga (1963) suggested its possible origin was from 'Long Orange'. The 'Long Orange' or 'Long Red' type carrot was mentioned in France in 1775, while the orange English cultivar 'Altringham' was first mentioned in 1834. Interestingly, the first white-rooted carrot was also described in Holland about the same time as the first orange types, thus suggesting a Dutch origin of this colour class as well.

Because it is believed that wild carrots have occurred in Holland since prehistoric times, some taxonomists have proposed that orange root colour variants resulted from hybridization of *D. carota* ssp. *carota* with *D. carota* ssp. *maximus* in the Mediterranean, and not from selection in Holland. Therefore, hybridization probably provided an opportunity for introgression of genes from wild carrot into cultivated carrot throughout its history of cultivation in Mediterranean and northern Europe regions. Heywood (1983) suggested a combination of hybridization with wild carrot and selection as the most likely origin of orange cultivated carrots. The free hybridization and geographic co-occurrence of wild and cultivated carrots means that hybridization is inevitable and, therefore, selection must be exercised to maintain necessary attributes for cultivation.

Small (1978) provided further insights into the origins of cultivated carrots. Based on numerical taxonomic analysis he found that the eastern and western cultivated carrots exhibited 'an uninterrupted spectrum of variants' with no clear line of demarcation. Wild carrots clearly differed from domesticated carrots and were further separable into separate but overlapping groups – *carota* (distinguishable most clearly by an inflexed or 'birds'-nest' mature umbel) and *gingidium* (flat or convex umbel). Small suggested that eastern carrots arose in southern Asia independently from western carrots and, as Banga proposed, white-rooted carrots arose fairly recently from western carrots. Heywood (1983) pointed out that the combination of both leaf and root differences between eastern and western carrots suggests that western carrots were not selected directly from eastern wild carrots but rather from hybrids between early European/Mediterranean carrots, white-rooted carrots and wild carrots.

Two questions remain concerning the origin of orange-rooted carrots. First, although hybridization with wild carrot has continued to occur since carrot breeding was undertaken, has wild carrot contributed alleles for orange root colour, or other important horticultural traits in subsequent selection? This question may be answered, as genes controlling root carotenoid accumulation are characterized molecularly so that their origin can be determined. Secondly, did the selection for orange root colour occur in northern Europe, or was this characteristic of western carrots selected in southern Europe, Asia Minor or even Afghanistan? The hybridization theory supposes that orange-rooted carrots occurred in the Mediterranean, particularly the Anatolian region of Turkey where, according to Mackevic (1932), referenced by Banga (1963), cultivated carrot diversity was especially great. Yet, as Banga noted, orange root colour was not specifically mentioned. Absence of this information is subject to personal interpretation so this question is likely to remain unanswered without new historical documentation. It should be mentioned that clearly orange-rooted carrots do occur in carrot germplasm from Turkey (Simon, 1996). Whether this characteristic was present in 12th century ancestors of this germplasm is not known.

Parsnip (*Pastinaca sativa* L.) is native to the region between the western Mediterranean and Caucasus mountains. Parsnips were known to the early Greek and Roman civilizations, but their spread into western Europe is not documented. Introduction to North America occurred in the early 17th century and slightly earlier to the West Indies. While an excellent root vegetable, its popularity seems to be limited to the UK, other north-eastern European countries, Canada and the USA. Selection enhanced root size, fleshiness and smoothness.

Celeriac (*Apium graveolens* L. var. *rapaceum* (Mill.) Gaudin. Celeriac and related celery and smallage are of the same species, are cross compatible, and thus share many similarities. Their origin is attributed to the eastern Mediterranean region. Selection and domestication altered the morphological appearance of celeriac compared to its relatives. Rather than development of thickened and succulent leaf petioles as with celery, and the leafy smallage, the enlarged hypocotyl and upper root of the celeriac plant is the edible product. Accordingly, the crop is often referred to as root celery or turnip-rooted celery. Celeriac is a relatively little known crop outside of northern and central Europe. Domestication furthered the plant's adaptation for production in cooler temperate regions.

Arracacha (*Arracacia xanthorrhiza* Bancroft) was domesticated in the Andean highlands of South America where it is of considerable importance to the people of those areas. Wild species are found in the Andes that most resemble the domesticated plants, whereas the greatest number of wild *Arracacia* species are found in Mexico. The often proposed day length and elevation requirements are not valid restrictions to arracacha production. Other than frost intolerance, production in other areas should not be limited, although cool to moderate temperatures (15–20°C) are preferred. Nevertheless, attempts at production in Europe and Asia met with no success. However, introduction into south-east Brazil resulted in a rapid expansion. Presently this is the major production region. In some locations its caloric and culinary value rivals potato in importance. Growers have over time improved the production of large, fleshy and smooth roots. Only recently has selection by plant breeders been directed toward improving the crop.

Much like parsley, **Hamburg parsley** (*Petroselinum crispum* (Mill) Nyman ex A.W. Hill var. *tuberosum*) (Bernh.) Crov.) had its origin in the Mediterranean and East Asia area, and similarly was domesticated in southern Europe. The domestication of Hamburg parsley was directed towards selection for large, fleshy and smooth roots, whereas parsley was selected for its aromatic and flavourful foliage. Hamburg parsley is of some importance in eastern Europe, especially during certain holiday periods.

Turnip-rooted chervil (*Chaerophyllum bulbosum* L.) is thought to have originated in the northern Europe region, the primary area of its present cultivation. Domestication probably emphasized improved root shape, fleshiness and taste.

Skirret (*Sium sisarum* L.) has an East Asia origin. The volume of its cultivation is limited. Domestication likely emphasized increased root size and aromatic flavour.

Asafetida (*Ferula assa-foetida* L.) is believed to have originated within the broad area that includes Iran, Turkestan, Afghanistan and India. Wild plants of the same genus, sometimes called common giant fennel, are found in central Asia. Any selection that might have occurred would be directed at improving the yield of the resinous juice extruded from plant tissues which is used for many flavouring purposes.

Greater burnet saxifrage (*Pimpinella major* (L.) Huds.) is found wild in most of Europe and parts of the Mediterranean and Asia Minor and is occasionally cultivated. Its strongly bitter-tasting root is used to flavour candies and liqueurs and is used in some pharmaceuticals. The related burnet saxifrage (*P. sasifrage*) is better known for it flavourful condiment seeds.

Great earthnut (*Bunium bulbocastanum* Koch) is of southern and temperate Asian origin. Great earthnut has tuberous roots that resemble and are used much like the edible roots of caraway, *Carum carvi*.

Epos (*Perideridia gairdneri* (Hook. & Arn.) Math.) is native to areas of the central and western USA and Canada. It is gathered from the wild and is seldom cultivated. Domestication efforts are in progress with the goal of increasing size and yield of the tuberous roots. The plant has many local names such as squawroot, yampah and edible-rooted caraway.

Cow parsnip (*Heracleum lanatum* Michx.), although also found growing wild in Siberia and Europe, probably is native to North America. Its parsnip-like storage root often is a subsistence food and sparingly used as a potherb. Folk medicinal use is made of roots and seeds.

Foliage crops

Many vegetable umbellifers are utilized mainly for their edible leaves, petioles and stems that are consumed fresh and as cooked potherbs. Celery and smallage crops are major foliage umbellifers and other relatively important crops include: parsley, cilantro (coriander) and Florence fennel. The production of additional foliage umbellifers is less extensive, but is significant in certain regions. Such vegetables include: salad chervil, culantro, Japanese hornwort, water dropwort, the pennyworts, lovages, angelicas, dill and several others. Some foliage umbellifers are used for garnish and for ornamental purposes.

Celery (*Apium graveolens* L. var. *dulce* (Mill.) Pwea.) and allied **Smallage** (*Apium graveolens* L. var. *secalinum* Alef.) were domesticated in the eastern

Mediterranean where they possibly originated. Although the Mediterranean basin also is often considered the centre of origin of the genus *Apium*, the wide distribution of some *Apium* species does not agree with this assumption. Wild forms of celery can be found in marshy areas throughout temperate Europe and western Asia and about 14 species of *Apium* are spread throughout the world from Australia and New Zealand to South America, as well as South Africa and the Mediterranean basin. *A. graveolens* is the only cultivated species in the genus.

Wild celery seed was used medicinally in ancient Egypt. It appears that celery was first cultivated about 400 BC as a medicinal plant in that part of the world and within the Roman Empire to combat various maladies, including drunkenness. Early literature mentions use of seed and seed oil for a variety of medicinal uses. Iranians used seed oil vapour to cure headaches. It has been claimed that beverages containing seed extracts were anti-flatulent, diuretic, nerve builders and aphrodisiacs. The Greeks produced a wine from celery called 'selinites'. Victory crowns made from celery foliage were awarded at athletic games on the Isthmus of Corinth, and wreaths of parsley at the Pan-Hellenic Games of Greece. Obviously the aromatic foliage of these plants was highly regarded.

The first culinary uses of celery took place in Europe, Egypt and China as potherbs, condiments and garnish. Most likely plants used for these purposes were leafy types with narrow non-succulent petioles typical of land races in the Mediterranean region, and likely mostly resembled present-day smallage rather than celery.

Celery cultivation was first reported in France about 1623, and as a crop for flavouring purposes in England about 1776. It is believed that celery was domesticated in the 16th century as a vegetable in Italy and France when milder-tasting forms appeared. In the USA cultivation of stalk celery first occurred in Michigan about 1874 and in Florida about 1897.

Important changes in plant structure occurred during domestication, these being selection for solid and succulent petioles rather than the enlarged fleshy hypocotyl root axis that constitutes celeriac (Vilmorin, 1950). Smallage, has a long history of cultivation in China, beginning about the 6th century, and its domestication apparently also occurred in Asia. Of the botanical varieties of *Apium graveolens*, smallage has the broadest climatic adaptation which is largely responsible for its wide distribution and extensive cultivation in Asia and many other locations. On the other hand, celery production is primarily located in temperate regions of North America and Europe.

Parsley (*Petroselinum crispum* (Mill.) Nyman ex A.W. Hill) is believed to have originated in the southern European and western Asian areas surrounding the Mediterranean basin. There it was cultivated more than 2000 years ago by the ancient Greeks, Romans and other civilizations. Its culture spread into western Europe during the 15th and 16th centuries and then to other temperate regions worldwide, and parsley is even grown in many high elevation subtropical areas. Domestication emphasized high levels of leaf essential oils for enhanced flavour and aroma, whereas the botanical variety, *P. crispum* var. *tuberosum*, was selected for large, fleshy and smooth roots.

Cilantro (*Coriandrum sativium* L.) origins are uncertain, but most indications suggest the Near East (Diederichsen, 1996). However, this does not completely coincide with centres of variation since distinct forms of coriander can be found in India, Central Asia, the Near East and Ethiopia. The spread of coriander through the Old World at an early date likely resulted in different plant forms. Coriander was discovered in Egyptian tombs dating back to about 900–1100 BC. Other indications suggest an earlier presence, as much as 2500 BC. According to Sanskrit writings, coriander may have been utilized as far back as 5000 BC. One report of fruit found in a cave in Israel was dated back to 6000 BC. The Romans were responsible for the movement of coriander into northern European regions. Its introduction into China may have occurred well before the 5th century.

The cultivated plant has many traits of wild coriander, but selection has resulted in larger and more abundant foliage types for vegetable use, and other types of coriander with higher yields of seeds containing increased levels of essential oil. Cilantro has considerable importance in New and Old World regions. The crop is very popular in Central and South American countries, and throughout South-east Asia and China. Although cilantro production is substantial, accurate production statistics are not easily obtained.

Florence fennel (*Foeniculum vulgare* Mill. var. *azoricum* Thell.) is native to Mediterranean regions of southern Europe and western Asia. This crop continues to be a popular and versatile vegetable in those and nearby regions. Its history of cultivation is old since it is known to have been a vegetable appreciated by the ancient Greek and Roman civilizations. Improvements since that period enlarged the size and fleshiness of the edible petiole bases and the aromatic content of foliage and seeds.

Chervil (*Anthriscus cerefolium* (L.) Hoffm.), also known as salad or garden chervil, as well as French parsley, is native to the region of south-east Europe and south-western Asia. The Romans introduced the plant into Europe and Britain. Its naturalization occurred in Europe and North America. The major area of production is Europe, and the crop is especially popular in France. Selection has enhanced the flavour of the foliage.

Culantro (*Eryngium foetidum* L.) has a tropical America and West Indies origin, and is a traditional potherb crop in those areas and nearby Central and South American countries. With its introduction to the Orient, it has also achieved considerable popularity in China, South-east Asia and Indonesia. The edible foliage has a strong aromatic and flavour resemblance to cilantro. Selection is aimed at improved plant leafiness, flavour and yield.

Japanese hornwort (*Crytotaenia japonica* Hassk.) is believed to have originated in eastern Asia and Japan. Wild forms are found in China, Vietnam and Japan. It is widely grown as a popular potherb in the Orient, South-east Asia and Indonesia. The crop has many common names such as mitsuba,

Japanese chervil and Japanese parsley. Domestication improved foliage size, stem succulence and yield. A similar plant of North American origin, *Crytotaenia canadensis* var. *japonica* is known as hornwort, white or wild chervil.

Water dropwort (*Oenanthe javanica* (Blume) DC.) is probably of Asian origin since wild forms of this semi-aquatic and aquatic perennial are found in freshwater marshes and along streams in Japan, East and South-east Asia, and Oceania. The first report of its cultivation was in Japan in the 10th century, although wild forms were previously used in Japan, China and other parts of Asia. Domestication improved petiole thickness and leaf yield.

Asiatic pennywort (*Centella asiatica* (L.) Urban.), despite its Asiatic name, is a perennial indigenous to the south-eastern USA. Asiatic pennywort is a popular salad and potherb vegetable in many Asian countries. A creeping and a bush-like form are cultivated. The latter is better suited for multiple harvesting and accordingly yields more. The creeping types are more often harvested from uncultivated growth.

Lawn pennywort (*Hydrocotyle sibthorpioides* Lamk.) is of South-east Asian origin. It is a weedy plant, frequently found in lawns in tropical and many subtropical areas. Its production for potherb use remains limited, as it is sparingly cultivated, but is frequently gathered from unattended growth.

Lovage (*Levisticum officinale* W.D.J. Koch). The alpine regions of Italy and France demarcate the origins of lovage. The plant has a long history of use and was known to the ancient Greeks and Romans. Since the Middle Ages, the crop has been a popular potherb as well as a flavouring and folk medicine product in northern and central Europe. The area of its cultivation for vegetable potherb use is relatively limited, although lovage continues to have some popularity in several European countries.

Scotch lovage (*Ligusticum scoticum* L.) is distantly related to lovage, but is native to the coastal regions of north-west Europe and England. Yet another lovage is **black lovage** (*Smyrnium olusatrum* L.), also known as Alexanders or horse parsley, which probably is native to southern Europe and the Balkan peninsula region. Black lovage has been mentioned in ancient Greek and Roman writings, and was naturalized throughout Europe following its introduction by the Romans. Black lovage has largely been replaced by celery which it resembles in appearance, use and flavour. Accordingly, its cultivation as with other lovages is minimal.

Angelica (*Angelica archangelica* L.), also known as garden angelica or European angelica, is native to Eurasia and was naturalized in Europe. Its use has been recognized for more than 1000 years. All plant parts can be utilized for food and flavouring. Selection likely was directed to improve stem and petiole size and tenderness, and seed oil yield. Commercial production is conducted in several northern European countries. Although the plant is known worldwide its cultivation in other areas is mostly limited to home gardening.

American or purple angelica (*Angelica atropurpurea* L.) has a North American origin centred about south-east Canada and the north-eastern and north central area of the USA. Often used from the wild, the plant occasionally is cultivated.

Dill (*Anethum graveolens* L.) is native to southern Europe and western Asia. Dill has an ancient history of food flavouring and medicinal uses since 3000 BC which involved the use of foliage (dill weed) and seeds. Dill was well known in ancient Egypt, in the Greek and Roman civilizations, and has been cultivated in Europe for many centuries. Presently dill is produced in many countries but the major volume of production, mostly as a condiment seed crop, occurs in the Indian sub-continent, Europe and the USA.

Bishop's weed (*Aegopodium podagraria* L.) is mostly a weedy plant and is seldom cultivated. Its probable origin was in Eurasia. Known for many centuries for its potherb and folk medicine qualities, its limited popularity is mostly confined to European countries.

Samphire (*Crithmum maritimum* L.), also known as rock samphire and sea fennel, is believed to have originated along rocky coastal regions of Europe. Not surprisingly, the plants are salt tolerant. Gathered from the wild or sparingly cultivated for its succulent shoots and leaves, the plant is not widely known or used.

Sea holly (*Eryngium maritimum* L.) also originated in coastal areas of Europe, although its naturalization occurred in the Atlantic coastal area of the USA. Young tender shoots are consumed. However, the crop is of little consequence, and thus receives little attention for improvement.

Seed crops

Several umbellifers are grown primarily for their seeds which contain specific essential oils and other compounds used for condiment and flavouring purposes. Additionally, some crops primarily grown for their seeds also have secondary uses of their foliage as flavouring for various dishes, soups, salads, as garnish and occasionally for decorative purposes. On the other hand, some foliage umbellifer crops also have a secondary use because of their aromatic seeds. These include: celery, Florence fennel, dill, angelica, coriander, lovage, chervil and sweet cicely. The best known plants grown mainly for their seeds are: coriander, caraway, fennel, anise, Indian dill, cumin, black cumin, burnet saxifrage and ajowan. Seed extracts from many of these seed crops are used in herbal medicines.

Coriander (*Coriandrum sativum* L.) probably originated in the Near East. In terms of volume, the coriander seed crop is much more important than cilantro, the vegetative product (foliage) of the same plant. Seed production is carried out in Europe, mostly eastern European countries and Asia Minor, also Morocco, India, Romania and Argentina. Diederichsen (1996) indicates that over 500,000 ha are involved in the production of coriander seed.

Caraway (*Carum carvi* L.) appears to have had its origin in Asia Minor.

Evidence of caraway was found in middle eastern Asia about 5000 years ago. The plant was well known to the ancient Egyptians. Its introduction from northern Africa into Europe occurred about 1000 years ago. Naturalization occurred mostly in northern Europe and North America with crop improvement directed toward increased seed size and essential oil content. Production of caraway seed is significant in northern Europe, namely The Netherlands, Canada, the USA, Scandinavia, Russia and other temperate countries. The tuberous roots of caraway are edible and somewhat popular, especially in China.

Fennel (*Foeniculum vulgare* Mill.) and **Sweet fennel** (*Foeniculum vulgare* Mill. var. *dulce* Batt. & Trab.) are known as bitter fennel, and sweet fennel or sweet anise, respectively, and are found wild in south-west Europe and north-west Africa where they have been cultivated for thousands of years. Their seeds and expressed essential oils are used primarily as condiments and seasonings. Major producers of fennel seeds are India and China, with significant additional production in Spain, France, Italy, other European countries, Egypt, Argentina, Indonesia, Japan and the USA. The seeds of another plant, *Ferula communis* L., known as common giant fennel, are also used for culinary seasoning. The plant is infrequently cultivated and often is a weed in the same European and African locations where wild fennel is found.

Anise (*Pimpinella anisum* L.) is reported to be indigenous to areas of Eurasia and Africa surrounding the eastern Mediterranean basin, although some botanists specify Egypt as a centre of origin. Anise seed was a source of flavouring for ancient Rome. It is widely cultivated, but not intensively, throughout the world. Most production is centred around the Mediterranean region, and also in Turkey, Ukraine, Moldavia, Georgia, India and Mexico. Through continued selection seed yield and aromatic compound content were increased.

Dill – see preceding foliage crops discussion.

Indian dill (*Anethum sowa* DC.) has its probable origin in India where its cultivation is extensive. Significant production also occurs in Japan and Indonesia. Selection for crop improvement emphasized increased seed size, essential oil content and yield.

Cumin (*Cuminum cyminum* L.), as a native of the eastern Mediterranean and northern Africa region, has an ancient history of use as far back as 5000 BC. Production continues to occur in these areas, but the major production, much of which is exported, occurs in India and Iran. Domestication concentrated on increasing seed yield and the content of its flavouring components.

Black cumin (*Bunium persicum* (Boiss.) Fedtsch.) is a non-cultivated plant often found in areas of the Middle East, especially Iran. It produces smaller and sweeter-tasting seeds much like those of cumin and is similarly used. This plant should not be confused with *Nigella sativa*, a member of the *Ranunculaceae* family which also has an origin in the Mediterranean area and is also known as black cumin or Roman coriander, and whose seeds are also used for seasoning.

Sweet cicely (*Myrrhis odourata* (L.) Scop.). The mountainous areas of Asia Minor and Europe are thought to be the regions of origin for sweet cicely. Its flavoursome seeds, foliage and roots have been used since the ancient Roman period. Domestication improved yield and the quality of all edible plant portions.

Burnet saxifrage (*Pimpinella saxifrage* L.) is found wild in areas of the UK and Ireland, in much of Europe, and parts of the Mediterranean and Asia Minor. Limited cultivation occurs in Mediterranean areas, northern Europe, and also in Asia Minor. Although the essential oils in the seed used in flavouring are the primary product, the young leaves occasionally are used in salads.

Ajowan (*Trachyspermum ammi* (L.) Spr. ex Turr.), apparently native to Egypt and Asia Minor, is cultivated in that region and in large volume in India mainly for condiment use of its seed. However, young seedlings are used as a potherb.

WORLD IMPORTANCE

With the exception of carrots and possibly celery and smallage, few of the umbelliferous vegetables can be considered as major food sources. Among the vegetable umbellifers, carrots are the major contributor of calories, although consumption of arracacha is a significant part of the daily caloric intake of some populations in Brazil and countries of the Andean mountain region. Nevertheless, in the aggregate, the vegetable umbellifers are important contributors to world food consumption because of their diverse and unique flavouring and textural characteristics that complement many diets. Furthermore, their value becomes more significant when the favourable nutritional qualities of the many vegetable umbellifers are considered. Vegetable usage of umbellifers varies with their availability and the food habits of different cultures.

Utilization of the various umbellifer crops often is fairly regional. For some populations, items such as cilantro are almost a daily essential in many daily food preparations. Consumption of parsnips and Hamburg parsley is largely centred in Europe. Florence fennel is most often used in southern and central Europe; chervil has its major popularity in France; water dropwort and Japanese hornwort consumption occurs primarily in eastern and south-eastern Asia. The lesser known umbellifers have limited production and correspondingly local and limited utilization.

In recent decades, the increased consumption of several vegetable umbellifers, carrots in particular, is due in part to a heightened awareness of the health-contributing attributes of some of these plants. A continuing increase in health consciousness should favour continued preferential expansion of their consumption. The development of cultivars better adapted to non-temperate climates should improve production and yields. Additionally, consumption has increased because of the development of many new cultivars that have provided improved flavour. Improved handling practices, expanded refrigerated storage, and other postharvest technologies have and will further improve quality retention and thus stimulate greater utilization.

Distribution, production areas and statistics

Many of the vegetable umbellifers fall into the category of cool-season crops, and thus their production most often occurs in temperate latitudes. Nevertheless, significant production does occur in some subtropical and some tropical latitudes, primarily at high altitude locations.

Presently, world production statistics from the United Nations Food and Agriculture Organization (FAO) are only reported for carrot. These statistics only include carrots cultivated for human consumption, and not production in small gardens or for animal feed.

For the 1997 period reported (Table 1.3) almost 38% of world carrot production occurred in Asian countries, followed by the European and North

Table 1.3. World production of carrots, 1997.

	Area (ha × 10^3)	Yield (t ha^{-1})	Production (tonnes × 10^3)
World	834	21.54	17,977
Europe	314	20.74	6805
Asia	327	20.86	6815
North and Central America	69	33.5	2327
Africa	69	12.49	863
South America	45	19.56	876
Oceania	10	30.99	311
Leading countries			
China	203*	22.15	4506*
USA	48	41.80	2024
Russian Federation	125*	12.00	1500*
Poland	30	26.96	799
Japan	26*	29.02	740*
France	17*	37.47	640*
United Kingdom	15*	42.57	630*
Netherlands	8*	51.19	430*
Italy	10	41.41	408
Canada	9*	38.62	331*
Ukraine	30*	10.83	325*
India	22*	14.55	320*
Germany	8	40.25	313
Spain	7*	46.15	300*

*Estimated.
Source: 1997 *FAO Production Yearbook*, Vol. 51, FAO, Rome, 1998.

and Central American regions at 36% and 14%, respectively. Africa and South America (less than 5%) were well behind, and the Oceania region, namely Australia and New Zealand, were responsible for about 1.7% of total world production. Morocco, Nigeria, Algeria, Egypt and South Africa are significant African producers, and Venezuela, Argentina, Colombia and Peru are important South American producers. Indonesia, Turkey, Australia and Mexico, each with more than 200,000 t, are other important countries in their respective regions.

Annual world production of carrots increased from 5.5 to more than 16 million tonnes (Mt) from the period 1963 to 1996 (Table 1.4). Percentage-wise the rate of gain was most notably in Asia. African and South American production has also steadily increased during the past 30 years. Overall, the rate of increase was greater than the world's population growth rate, and the overall increase in total world vegetable production. Europe has long been the major world production region. However, recent Asian production (primarily China) has exhibited a very rapid rate of increase, and in 1997 displaced Europe as the leading production regions.

Per capita carrot consumption in the USA has more than doubled in the last 30 years, with a considerable gain in both fresh and frozen forms (Table 1.5). During the same 30 year period, celery production and consumption was relatively stable.

Much of the increased carrot production was due to the increase in production area rather than higher yields. World carrot yields have been relatively consistent ranging annually between 21 and 25 t ha^{-1}. Presently, the world volume of carrots produced is about 1% of the total of all world root, tuber and other vegetable crop production. Current production statistics for vegetable umbellifers and condiment crops other than carrot and celery are difficult to obtain or are not available because their limited scale of production and utilization often are not reported.

In summary, the more extensively grown root vegetable umbellifers are: carrot, parsnips, celeriac and arracacha. Carrots have nearly worldwide distribution, which is not the case for parsnips, celeriac or arracacha. Parsnip and celeriac have limited popularity and are mainly grown and consumed as winter vegetables in northern and western Europe due in part to their good storage characteristics. Arracacha is essentially limited to the Andean regions of South America, several countries of Central America and south-eastern Brazil. In Brazil, which is the major producer, the crop has become a fairly popular vegetable. Other important producers are Colombia, Ecuador and Venezuela. World production of arracacha is estimated to involve more than 30,000 ha (Hermann, 1997). Hamburg parsley and turnip-rooted chervil are root vegetable umbellifers of moderate production volume. Both are favoured mostly by central and western European populations. Skirret is a minor root crop having some popularity in eastern Asia.

Major foliage umbellifers representing relatively large volume for

Table 1.4. World carrot production, 1963–1995.

	World		Africa		North and Central America		South America		Asia		Europe		Oceania	
Year*	Area	Prod	Area	Prod	Area	Prod	Area	Prod	Area	Prod	Area	Prod	Area	Prod
1963	189	4226	15	165	39	922	13	167	16	447	103	2443	3	82
1970	415	7908	19	214	41	1099	19	281	116	1776	118	3049	4	117
1975	459	8747	25	291	42	1241	21	322	125	2078	120	2906	4	126
1980	515	10,499	31	383	46	1328	27	465	142	2677	129	3591	5	141
1985	558	12,156	35	416	53	1480	30	545	163	3369	134	4043	5	158
1990	617	13,696	43	558	56	1773	34	630	181	4003	143	4304	6	193
1995†	756	16,062	64	776	65	2161	36	689	256	5575	327	6615	8	246

*Values are the average of the preceding 3 years with the area as (1000 ha) and production as (1000 Mt).
†Former USSR republics included in appropriate European or Asian grouping.
Source: *FAO Production Yearbook* Statistics. China and Asia USSR not included in 1983 estimate; Nigeria not included before 1985.

Table 1.5. US carrot and celery production and per capita consumption, 1925–1995.

	Carrot						Celery		
	Area	Production	Consumption (kg year^{-1} capita^{-1})				Area	Production	Consumption
Year*	(1000 ha)	(1000 Mt)	Fresh	Canned	Frozen	Total	(1000 ha)	(1000 Mt)	Fresh (kg year^{-1} capita^{-1})
1925	6	100	2.0	—	—	2.0	10	258	3.1
1930	14	258	3.1	—	—	3.1	14	421	4.0
1935	17	299	3.5	0.1	—	3.6	15	390	3.8
1940	25	439	3.2	0.1	—	3.3	17	476	3.6
1945	38	738	4.7	0.2	—	4.9	18	530	3.7
1950	35	675	3.8	0.2	—	4.0	15	607	3.9
1955	32	698	3.5	0.2	0.1	3.8	14	702	3.9
1960	32	747	3.2	0.2	0.2	3.6	14	689	3.6
1965	33	770	3.0	0.3	0.2	3.5	13	639	3.1
1970	31	834	2.7	0.3	0.3	3.3	13	711	3.3
1975	30	924	3.0	0.3	0.4	3.7	13	743	3.3
1980	34	978	3.1	0.2	0.5	3.8	15	838	3.4
1985	36	1060	3.5	0.4	1.1	5.0	14	824	3.3
1990	40	1332	3.5	0.4	1.2	5.1	14	899	3.3
1995	48	1732	3.8	0.5	1.3	5.6	10	750	2.8

*Values are the average of preceding 3 years.

Source: USDA Agricultural Statistics; canned and frozen per capita values given on processing weight basis before 1985, on farm weight basis after 1980.

commercial production with well-defined markets are: celery and smallage, parsley, cilantro and Florence fennel.

Celery is a significant crop mostly in North America and in Europe, and because of its increasing popularity its production has greatly expanded into other areas. Several countries produce winter crops of celery for export into European markets. Celery is produced from about 10,000 ha annually in the USA. On the other hand, smallage production is very extensive, being considerably more than that of celery. In much of Asia, especially China, smallage is a very common and frequently used salad, potherb and food-flavouring ingredient. Parsley also has a wide world distribution and many hectares are dedicated to its production for fresh and processing use. The major production and consumption of Florence fennel occurs in Europe. Its popularity is greatest in Italy, which alone is reported to cultivate more than 20,000 ha annually. Spain and France are other large producers. A small volume of several crops, such as celery, Florence fennel, parsley, as well as carrots and celeriac, are grown in glasshouses or similar plant-growing facilities.

The principle umbellifer condiment seed crops grown in commercial-scale operations are: coriander (> 500,000 ha), caraway (> 100,000 ha), anise, cumin (200,000 ha with 80,000 ha in India alone), dill, fennel, Indian dill and sweet cicely. The international trade for these condiment seed crops is very large and important. Celery seed production exceeds 2000 t annually.

The extent of the production of condiment seed crops is difficult to ascertain, because the information is often proprietary. While much of the umbellifer seed crop production is grown on a contracted basis, much is also produced for conventional marketing, and a large volume is also produced that does not enter market channels.

Dill is grown in the UK, Hungary, Germany, other European and several former USSR republics, and India. Interestingly Indian dill, widely grown in India where it is a common winter crop, is also grown to considerable extent in Japan. Cumin is grown in India, Iran, Egypt, Turkey, Morocco, Azerbaijan, Kazakstan, Moldavia, Turkestan, Ukraine and Uzbekistan.

International trading is important, but except for regional distribution the exportation of vegetable umbels is rather limited mostly due to their bulk, value and/or perishability. The notable exception is for condiment seeds, flavouring products and some processed products. Production of many minor crops occurs largely on a home garden scale for producer use and/or for limited local marketing. These may include: angelica, Asiatic pennywort, black cumin, different lovage species, burnet saxifrage, culantro, asafetida and ajowan. Quantification of the volume and consumption of another grouping of edible umbellifers such as cow parsnip, common cow parsnip, epos, great earthnut, Bishop's weed, cow parsley, lawn pennywort, samphire, sea holly and common giant fennel is unavailable because they are seldom or never cultivated.

2

Botany and Taxonomy

GENERAL BOTANY

The most easily recognized characteristic of the *Umbelliferae* is their compound umbel (umbrella-like) inflorescence (Fig. 2.1a). The flowers occur in umbellets where pedicels of each flower radiate from a common point. The umbellets in turn arise on pedicel rays originating from the apex of the inflorescence stalk (Fig. 2.1b). Other common characteristics are alternate compound leaves, and a strong distinctive aroma, due to essential oils and other compounds unique for each species. Flowers are small, usually bisexual with a five-lobed calyx, five petals and stamens, and a two-celled inferior ovary (Fig. 2.2a, b). The fruit are dry schizocarps consisting of two ribbed or winged mericarps that can separate at maturity, each of which are the true seed (Fig. 2.2c, d). Mericarps are small, longer than they are wide, and make up the longitudinal hemisphere of the fruit. The fruit, mericarps and other plant parts often have oil ducts.

Additional characteristics are the cotyledons which assume two distinct shapes: round as characteristic for celery, parsnip, anise and angelica; and long as is characteristic for carrot, coriander, cumin, dill and caraway. The true leaves are entire in some species, divided in others, or graduate from entire to very finely divided during development.

The extent of the variation regarding other plant parts is rather broad. For example, the enlarged fleshy storage taproots of carrot and parsnip, the lateral storage roots of arracacha, and the fleshy hypocotyl–root axis of celeriac differ from the non-storage roots of other species. Leaf petiole thickness, length and shape differ greatly, especially noticeable when the long thickened petioles of celery are compared with those of parsley. Another contrast is that of leaflet differences such as the finely divided leaflets of Florence fennel versus the considerably broader leaflets of crops such as celery or lovage.

(b)

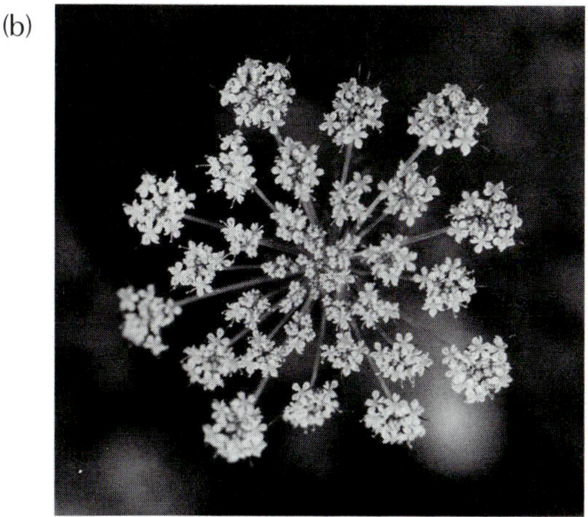

Fig. 2.1. (a) Representative carrot umbel and (b) top view of carrot umbellets.

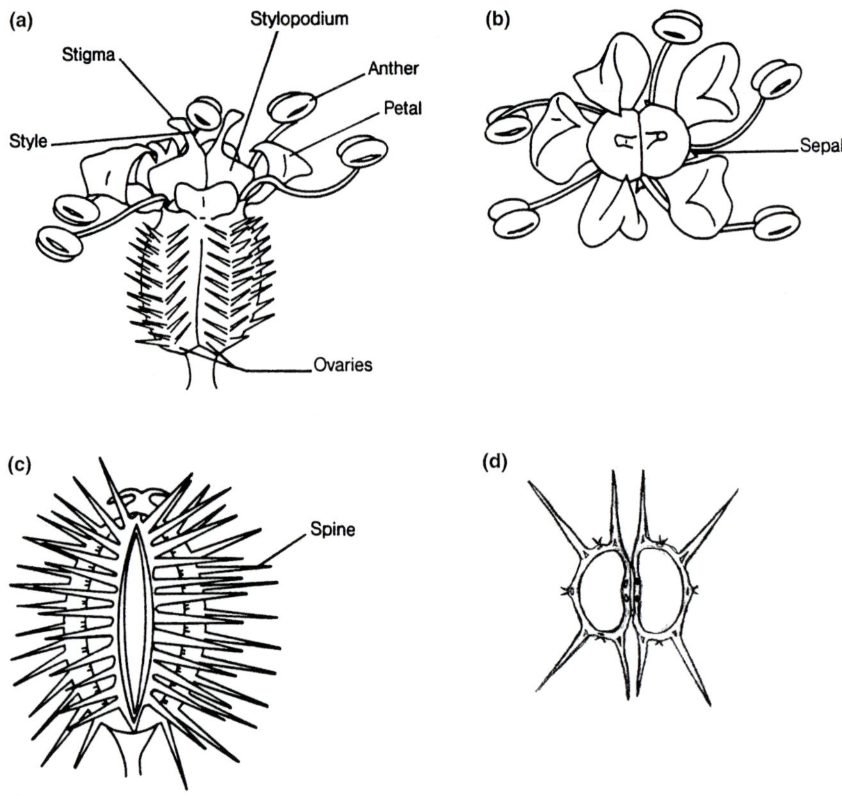

Fig. 2.2. Representation of carrot flower: (a) side view, (b) top view, (c) schizocarp (fruit) and (d) paired mericarps (seeds).

CARROT BOTANY

Hypocotyl and stems

The young carrot seedlings soon after emergence show a clear demarcation between the taproot and the hypocotyl. The latter at first is thicker and bears no lateral roots. The upper extremity of the hypocotyl, typically 20–30 mm in length, is terminated by the cotyledonary node. The hypocotyl gradually merges with the united bases of the cotyledons. The bases of the cotyledon are pushed apart as the axis that will produce the rosette of leaves develops from the plumule enclosed between the cotyledons. Subsequent leaves arise from stem nodes. The stem of the carrot plant during its vegetative state, as with many vegetable umbellifers, is just above ground and is greatly compressed so that the internodes are not clearly discernible. The stem apex is only slightly convex and remains so until seed stalk initiation. At that time the apex becomes narrow and pointed as it elongates and extends upward. Along with further elongation the stem produces a highly branched inflorescence. The stem and branches are rough and hairy and have enlarged nodes and hollow internodes. A plant may have one or several floral stalks developed which can range in height from 60 cm to as much as 200 cm.

Leaves

Following carrot seed germination, the emerging cotyledons elongate and spread apart. The first true leaf emerges after 10–15 days with a new leaf developing at about the same interval throughout most of the active growth. Leaves of the basal rosette are alternate and compound. Petiole bases are expanded and sheath-like at their basal attachment. The new leaves develops centripetally in a spiral within the basal clasping of preceding petioles. Leaf blades are two to three pinnate, the leaflets being repeatedly divided (pinnatifid) with small highly lobed segments that are oblong to linear and acute (Fig. 2.3b).

Roots

Anatomically, most of the storage root as shown in (Fig. 2.4) is comprised of parenchymatous phloem and xylem permeated by vascular tissues with cambium sections joining together in a cylinder. The ontogeny of the hypocotyl and root have been well detailed by Havis (1939) and Esau (1940). High-quality carrots are those having a large content of phloem relative to that of the xylem ('core') portion. Although a 'coreless' carrot is not possible, some cultivars have a small xylem region. When the xylem tissues are similar in colour intensity to the phloem such roots can appear to be coreless.

The storage root shape of many carrot cultivars is conical, but the extent of taper varies between cultivar types. Some cultivar types are cylindrical, round or various intermediates of these shapes. For some cultivars, root diameter may be only 1 or 2 cm, whereas, for others, the diameter may be 10 cm at the widest portion. Root lengths range from 5 to more than 50 cm in length, although the root length of most cultivars ranges between 10 and 25 cm. In addition to orange roots, yellow-, red-, purple- and white-fleshed roots are also grown. Alpha- and β-carotene are the major pigments responsible for orange and yellow roots. β-Carotene often may represent 50% or more of the total carotenoid content, and usually is about twice that of α-carotene. Carotenoid content is not

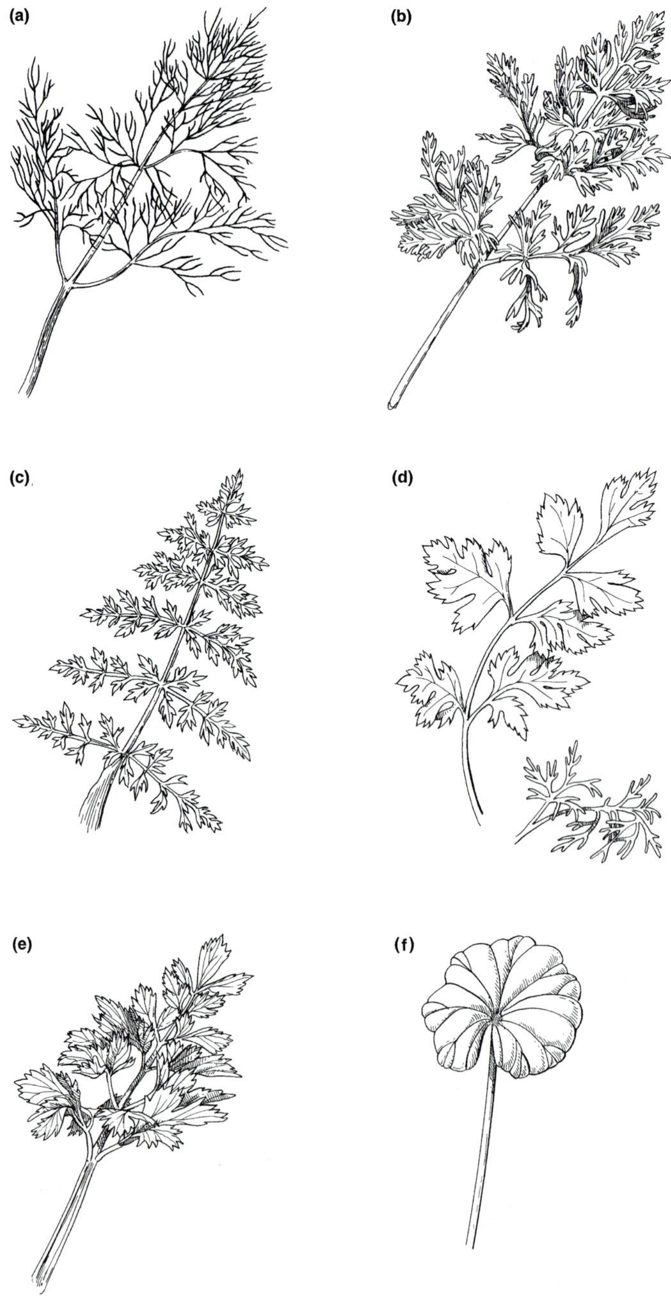

Fig. 2.3. Leaf shapes of several vegetable *Umbelliferae*. Foliage of various umbellifers exhibit a wide range of forms. They range from compound thread-like leaflets such as dill (a) to the circular non-segmented leaves of lawn pennywort (f). (a) Dill, (b) carrot, (c) caraway, (d) coriander, (e) lovage, and (f) lawn pennywort.

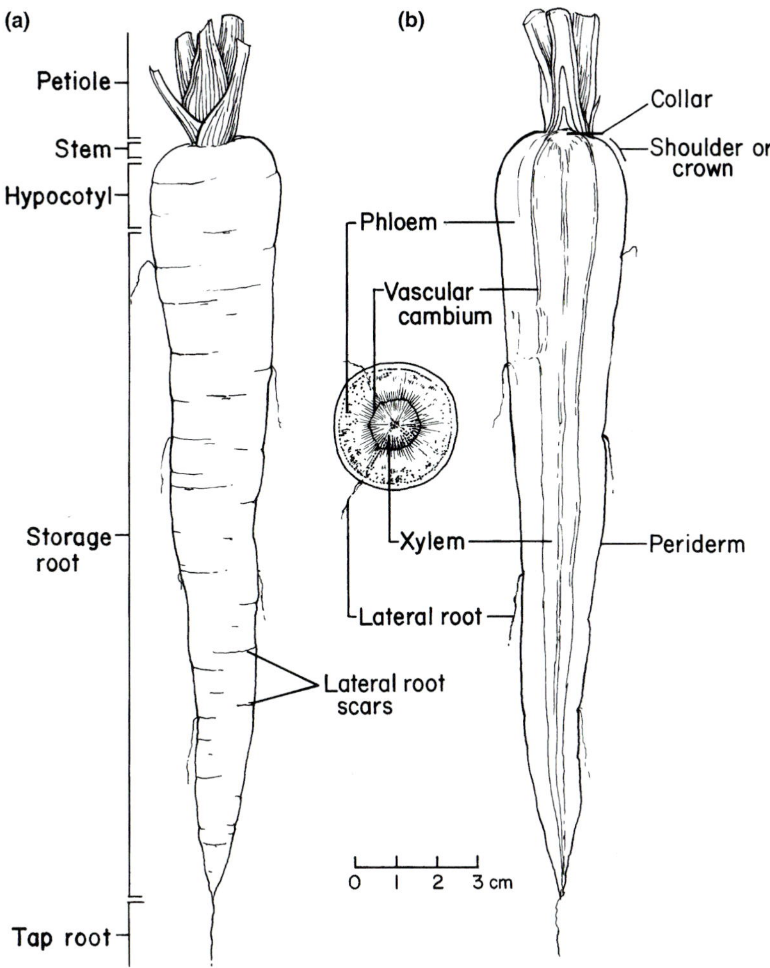

Fig. 2.4. Carrot storage root anatomy: (a) longitudinal and (b) cross-section.

uniformly distributed in the root because synthesis proceeds from proximal to distal tissues, and phloem tissues usually contain more carotenoids than the xylem. Xanthophylls are carotenoids largely responsible for yellow root colour, whereas lycopene determines red colour, and anthocyanins, not carotenoids, are responsible for purple root colour. White roots lack pigment.

Inflorescence, flowers and seeds

Floral initiation in carrot involves a morphological shift from the stem's relatively flat apical meristem producing leaves to an uplifted conical meristem capable of producing stem elongation and an inflorescence (Borthwick *et al.*, 1931). At first the floral axis grows only slightly, although with practice it can readily be identified visually. In the following weeks the hispid floral stalk greatly elongates and branches. The primary or 'king' umbel is produced at the terminus of the main floral stem (Fig. 2.5). Terminating the branches from the main stem are secondary umbels, and depending on further growth and stem branching, third, fourth and even higher order umbels can be formed. These progressively are smaller and develop later. A large primary umbel may contain as many as 50 umbellets, each with as many as 50 flowers (Fig. 2.1b). Umbels are subtended by a whorl of many long lobed bracts, and umbellets within each umbel by a whorl of entire or toothed lobed bracts.

Generally carrot flowers are perfect, small and white or occasionally greenish white or light yellow. Flowers consist of five petals, five stamens and an entire calyx (Fig. 2.2a, b). Typically anthers dehisce and stamens fall before the stigma becomes receptive. The anthers of the brown anther male sterile flowers degenerate and shrivel before anthesis. The other type of male sterile flower in which the stamens are replaced by petals is petaloid, and the petals persist. A simple method for estimating stigma receptivity is to look for a separation of the paired styles. The ovary is inferior and bilocular, each locule with a single functional ovule. On the upper surface of the carpels is a nectar-containing disc.

Floral development is protandrous and centripetal. Flowers usually first open at the periphery of the primary umbel. About a week later the process begins on the secondary umbels, to be followed a week or more later in the tertiary umbels. Still higher order umbels also follow this pattern. The flowering period of individual umbels usually ranges from 7 to 10 days. Thus a plant can be in the process of flowering for 30–50 days. The distinctive umbels and floral nectaries attract insects which are most responsible for performing pollinations. After fertilization and as seeds develop, the outer umbellets of an umbel bend inward causing the umbel shape to change from slightly convex or fairly flat to concave, and when cupped it resembles a bird's nest.

The fruit that develops is a schizocarp consisting of two mericarps, each mericarp being an achene or true seed (Fig. 2.2c). Upon drying the paired mericarps are easily separated. Premature separation (shattering) before harvest is undesirable because it can result in seed loss. Mature seeds are flattened on the commissural side that faced the septum of the ovary. The opposite side has five longitudinal ribs. Spines protrude from some ribs (Fig. 2.2d). These are removed usually by abrasion during milling and cleaning. Seed also contain oil ducts and canals. Seed size variation is common and can range from less than 500 to more than 1000 seeds per gramme (Fig. 2.6).

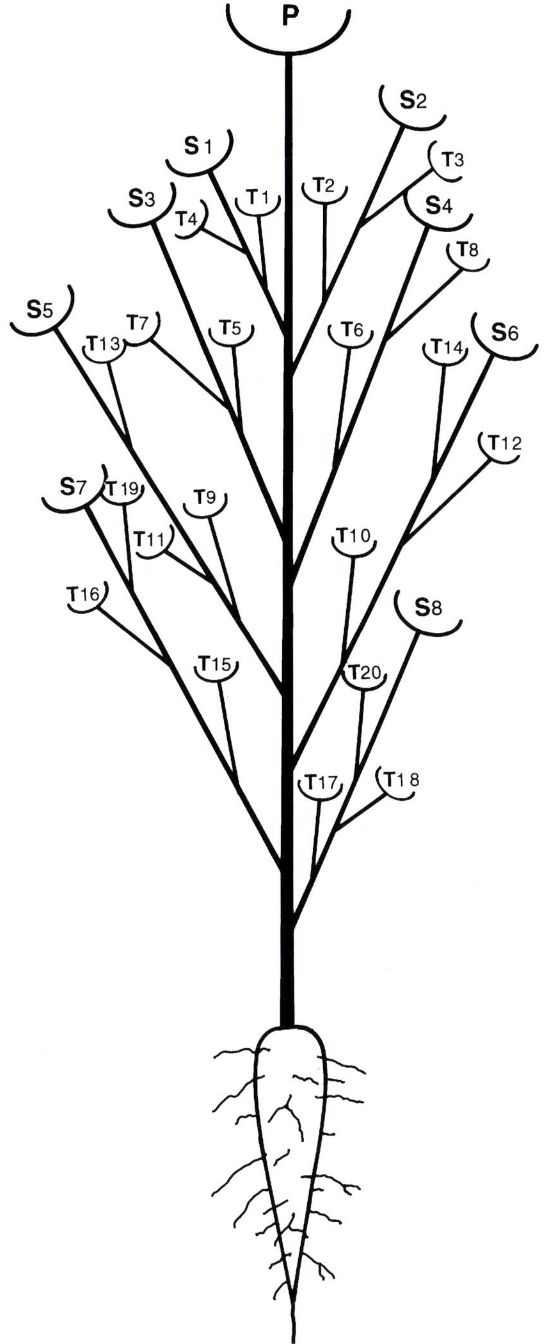

Fig. 2.5. Succession of umbel development in carrot. P = primary, S = secondary and T = tertiary; numbers indicate order of development of secondary and tertiary umbels. Fourth and higher orders may also occur.

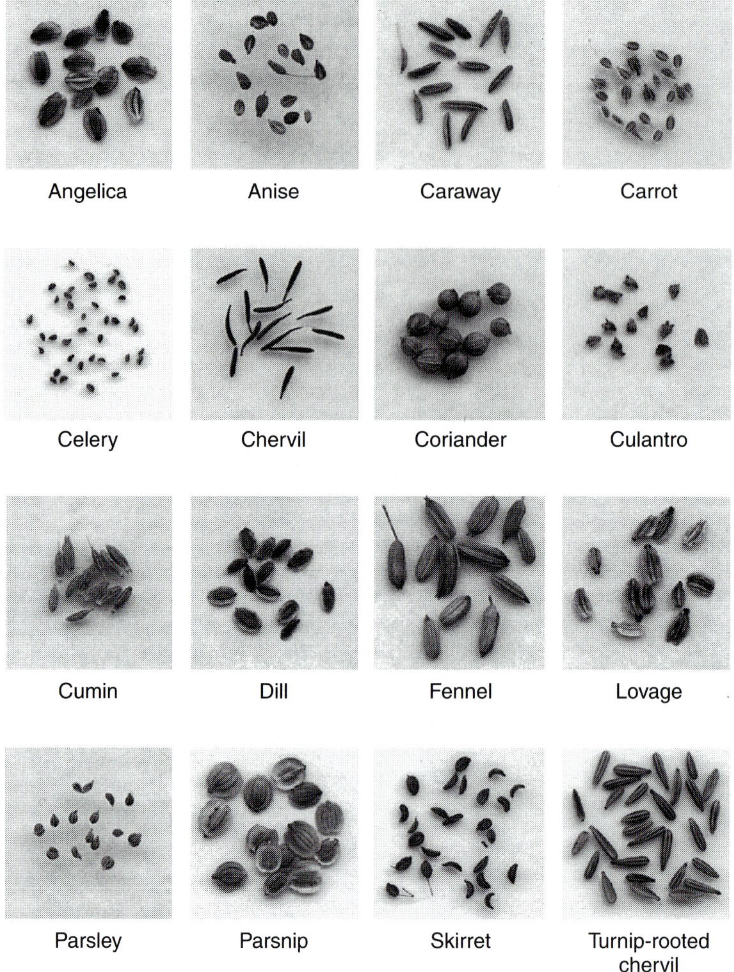

Fig. 2.6. Seed shapes of several vegetable *Umbelliferae*.

OTHER ROOT CROPS

Parsnip

Parsnips are biennial plants grown for their edible fleshy roots which, like the carrot, develop from tissues of the taproot and hypocotyl during the first year of growth. The long tapering storage taproot is usually smooth surfaced, but lateral roots, which detract from marketable appearance, occasionally are formed. Conical-shaped storage roots (Fig. 2.7) are most common, but some

Fig. 2.7. Parsnip roots presented for fresh market.

cultivars have bulbous-shaped storage roots, which tend to incur less breakage and are preferred by food processors for greater recovery of raw product. Parsnip root growth rate and development is relatively slow.

The apical meristem on the depressed short, compressed stem produces a rosette of many leaves. Parsnip leaves are pinnately compound, with the lower leaflets petiolate, the upper sessile and the terminal leaflets three-lobed. Leaves and leaf blade segments are much broader and larger than those of carrot. Petiole bases are slightly expanded and clasping. The floral stems are highly branched, grooved, hollow and capable of achieving a height of more than 150 cm.

Flowers without involucres occur in broad, compound umbels. Yellow-green flowers are epigynous, sepals are rudimentary and a nectary is present at the base of the two styles. Unlike carrot, the inflorescence does not have involucres. Flowering is protandrous, but pollination within the umbel is very common. The mericarps have two of their lateral ribs expanded into broad flat wings; seeds contain oil ducts. Seed size varies between 5 and

8 mm in length and about 275 usually weigh 1 g. Embryo development frequently is very slow and/or incomplete and thus low viability is often responsible for the notoriously slow and poor germination of parsnip seeds.

Celeriac

The edible product of the celeriac plant consists partly of the fleshy stem base, enlarged hypocotyl and enlarged taproot. Leaves develop as a rosette with long basal clasping petioles that crowns the hypocotyl. Celeriac leaf petioles differ from celery in that they are thinner, less solid and often hollow. Leaf growth tends to be shorter, less erect and exhibits a more open growth habit than that of celery, while the stem normally is less compressed. Much of the storage organ develops above the soil surface. Many well-developed lateral roots and a profuse mass of fibrous roots develop from the middle and distal portion of the taproot. This organ, often known as knob- or turnip-rooted celery, is generally globe-like in shape, with a light brown or tan exterior surface, and a white crispy-fleshed interior. During the reproductive state, stem elongation and inflorescence growth attains a height of 90–150 cm. Celeriac umbel, floral and seed characteristics are similar to those of celery and smallage.

Arracacha

Arracacha is a stout perennial plant grown for the production of its lateral fleshy storage roots (Fig. 2.8). These roots somewhat resemble those of carrot and parsnip with the interior tissues either white or yellow. Cultivars with white, yellow and purple exterior root colours are known, respectively, as blanca, amarilla and morada. Usually four to six lateral carrot size roots are formed clustered around a central rootstock which actually is the highly swollen stem; sometimes as many as 10 storage roots may develop. Root sizes range from 5 to 25 cm in length and between 2 and 6 cm in width. Individually these usually weigh between 100 and 300 g, although roots weighing 1 kg are occasionally obtained when storage root numbers are small.

Plant height ranges from 0.5 to 1.2 m. Foliage colour is green, often with a purple tinge, and occasionally deep purple. Leaves are broadly ovate, biternate or bipinnate, with leaflets ovate–lanceolate to ovate and 4–12 cm long. The leaf blades are acuminate, serrate and coarsely incised. Petiole length ranges from 8 to 45 cm and has a narrow sheathing. Petioles and the leaf blades resemble celery in appearance, taste and smell. The flowers are very small and are borne on relatively open compound umbels of 5–15 rays,

Fig. 2.8. Arracacha plant showing storage roots, stems or cormels and leaves. Normally, cormels are smaller in commercially harvested plants, which are grown 12 to 16 months. This specimen is 20 months old and was left to over-mature to illustrate its vegetative plant organs. Source: Michael Hermann, International Potato Center, Lima, Peru.

each supporting about 10–20 umbellets. Flowers are greenish when immature and become maroon or purple when flowers are mature. The styles are receptive several days before pollen is shed, and therefore out-crossing is promoted, although flowers are self-fertile. The oblong fruit are about 8–10 mm long and 2–3 mm broad, and prominently ribbed. During crop growth, removal of floral stems enhances root growth. If flowering is undisturbed, root yield is reduced, and edible quality decreases. Although seeds are produced, propagation is almost always made with offshoots, which normally form above or at the soil surface. In addition to storage root production, the edible celery-like petioles are often blanched by tying the leaves together for potherb use.

Hamburg parsley

The enlarged storage root of this cold-hardy biennial are harvested at the end of the first year's growth. The plants strongly resemble leaf parsley, especially the plain or flat leaf forms. Storage root enlargement is relatively slow and more similar to that of parsnip than carrot. Roots have a light brown exterior, are white fleshed, and aromatic. Root length ranges from 10 to 20 cm, and from 2 to 4 cm in width. A period of low temperature exposure shortly before or after harvest allows for the conversion of root starch to sugar, and thereby enhances root sweetness and flavour. Other plant characteristics are similar to those of the closely related leaf parsley.

Turnip-rooted chervil

Turnip-rooted chervil is not related and differs considerably from salad chervil. This hardy biennial is grown as an annual for its spindle-shaped tuberous roots. The 5–10 cm long roots have a grey-black exterior colour with the interior flesh yellowish white. Where temperatures are mild, the plants can be overwintered. Much like Hamburg parsley, the sweetness of the roots tends to increase with plant maturity and declining temperatures. Plants grow to a height of 90 cm. Leaves are decompound into lanceolate or linear segments, and are pubescent. Stems are grooved, purple and erect. Compound umbels bear small white flowers that produce 5–6 mm long, ovoid to linear-shaped fruit that have a striped appearance.

Skirret

Skirret is a perennial that is grown as an annual. Leaves are 60–90 cm long and bipinnate with one to three pairs of lanceolate toothed leaflets. Clusters of greyish, wrinkled, tuberous, white-fleshed roots are formed at the base of the stem (Fig. 2.9). Small white flowers are borne on the compound terminal inflorescence. The ovate-oblong fruit, about 3 mm long, are ribbed.

Asafetida

Asafetida are tall, foul-smelling perennial plants that range in height from 90 cm to as much as 3 m. Sulphurous compounds in the plant tissues are responsible for their offensive odour. Stout well-branched stems bear long leaves that closely sheath the stems. Basal leaves easily achieve 50–60 cm in length. The many clumped upper leaves have a noticeable odour and a cabbage-like appearance. The thick spindle-shaped roots are woody and

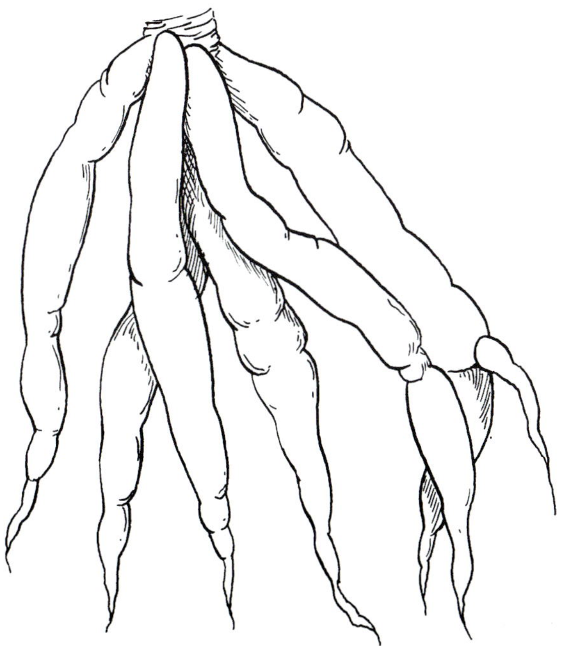

Fig. 2.9. Representation of clustered skirret roots.

contain a milky resinous juice, as do stems and all foliage. Cutting the root and basal stem tissues results in excretion of a milky resin at the cut surfaces. After it has dried, the resin is harvested for a variety of flavouring uses. Young shoots and leaves are used as potherbs in Central Asia. The clustered inflorescence bears greenish yellow flowers on nearly equal length rays. The seed of the flat-shaped fruit are about 10–12 mm in length. Several other *Ferula* species have similarities to asafetida growth and use.

Greater burnet saxifrage

Greater burnet saxifrage is a slender branching perennial usually growing from 30 to 45 cm in height, although some plants range up to 100 cm. The basal leaves are divided into many pairs of sessile bipinnate leaflets of variable leaf shapes that have a cucumber-like aroma. The stems are rough, and the thick roots, as much as 20 cm in length, have a very strong odour and bitter taste. White flowers are borne on flat-topped umbels. Fruit are shiny black and round or oblong; the seeds are about 2 mm in length.

Great earthnut

A perennial native to the Old World, great earthnut has relatively large thickened tuberous-like roots. Leaves are pinnately compound. Small whitish pink flowers produce a small compressed fruit.

Epos

Epos is a tall perennial reaching about 1–1.2 m in height that produces edible dark-skinned, nut-like, tuberous, white-fleshed storage roots. The large lower leaves have one or two pinnately compound leaflets that are linearly cut into short narrow segments. Upper leaves are usually simple. Compound umbels contain small white flowers. The nearly orbicular compressed fruit are 4–6 mm long.

Cow parsnip

Cow parsnip is a tall stout, boreal, spreading, pubescent perennial. Roots are long and relatively thick. Leaves are ternately compound with three to six ovate coarsely toothed leaflets about 5–10 cm long with wool-like under surfaces. White flowers bear small pubescent fruit.

FOLIAGE CROPS

Celery

Celery is typically cultivated as an annual plant, although most cultivars are biennials. After seed germination, the young growing seedling develops a strong spindle-shaped fleshy taproot which continues to enlarge during growth. The root system consisting of taproot and fibrous lateral roots may extend to depths beyond 60 cm. When the taproot is damaged or destroyed, as often occurs during transplanting, a well-developed system of fibrous, relatively shallow lateral roots originates from the remnant taproot. These profusely penetrate the soil from the surface usually to a depth of 20–30 cm with some deeper.

The celery plant is characterized by a rosette of mostly erect, succulent, relatively thick and solid petioles that are closely bunched to form what is sometimes referred to as a head or stalk. Petioles are long, broad, glabrous, crescentic in cross-section, with expanded sheathing and shingling bases. The tightly clustered petioles are attached to the flattened compressed stem. These fleshy petioles develop during the vegetative and first cycle of growth, during which time the stem remains compressed as a fleshy crown (Hayward, 1938).

The petioles consist mostly of large thin-walled parenchyma tissue and collenchyma strands associated with vascular bundles beneath the epidermis (Fig. 2.10). Ribbing on the abaxial (convex) surface is prominent because of the proximity of collenchyma tissues, whereas the adaxial (concave) surface is smooth. Vascular bundles and collenchyma strands cause celery petioles to be stringy.

Mature plants have from seven to 15 easily identifiable large petioles with additional progressively smaller petioles, usually hidden by the larger outermost petioles developing from the apical meristem. These innermost petioles, because of light exclusion, are light coloured and very tender. Collectively they are often referred to as the celery 'heart'.

Glabrous leaf blades on the long petioles are pinnate or bipinnate with apical pinnae having three sessile leaflets. Leaflet shapes are somewhat rhomboid, or almost triangular with bases cuneate, apex acute and margins dentate or incised (Fig. 2.10). Many celery cultivars have shiny dark green

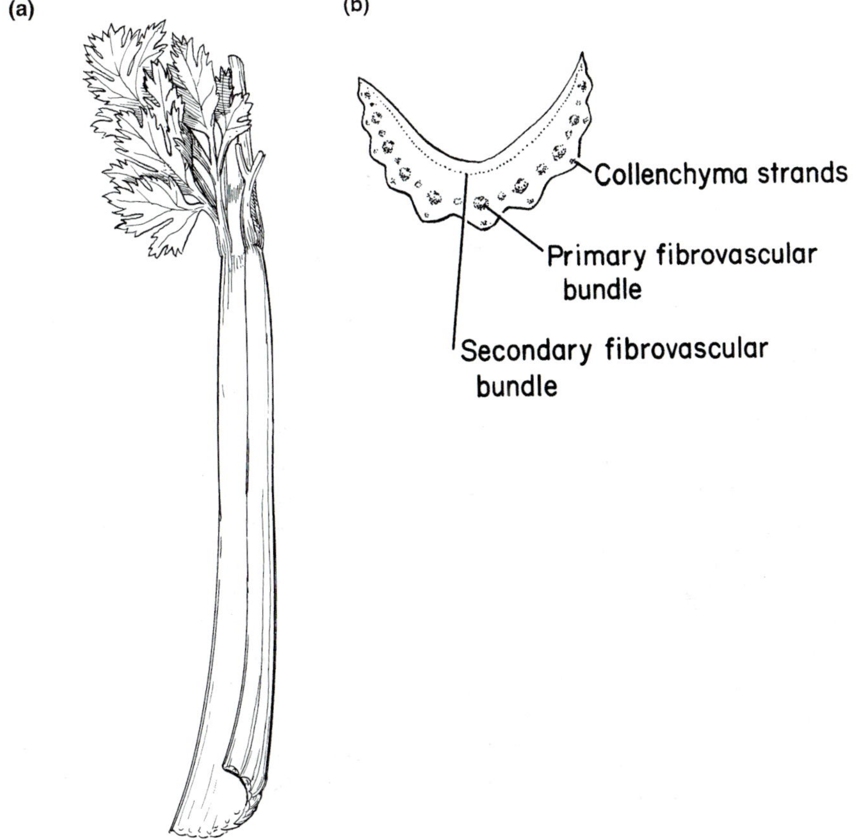

Fig. 2.10. (a) Celery petiole and (b) anatomy of celery petiole cross-section.

foliage, whereas self-blanching cultivars have light green or yellowish green leaves. Leaves of some celery cultivars have anthocyanin pigmentation, as do some cultivars of celeriac and smallage.

During the reproductive cycle of growth, the shoot apical meristem elongates to produce a seed stalk (Fig. 2.11b), which is branched, glabrous, strongly ridged and lignified. During such growth much of the upper portion of the terminal stem may become hollow. The small white or greenish white flowers of celery, celeriac and smallage are arranged in large compound umbels. The inflorescence differs from carrot in that the involucre and involucels are either absent or very small. Umbels are smaller and less compact. Flowering is protandrous with pollen usually released 3–6 days before stigma receptivity. The ovary is surmounted by a nectary disk and stylopodium supporting two short styles. The inferior ovary has two locules with a single ovule in each chamber.

The fruit, a very small flattened schizocarp, splits when mature into two single-seeded mericarps; but often remain attached by a filament called a carpophore. Mericarps have five primary corky ribs, are mostly oval-shaped and are very small; about 2500 seeds weigh 1 g. The seeds have abundant endosperm, a small embryo and essential oil-containing ducts, called vittae (Hayward, 1938).

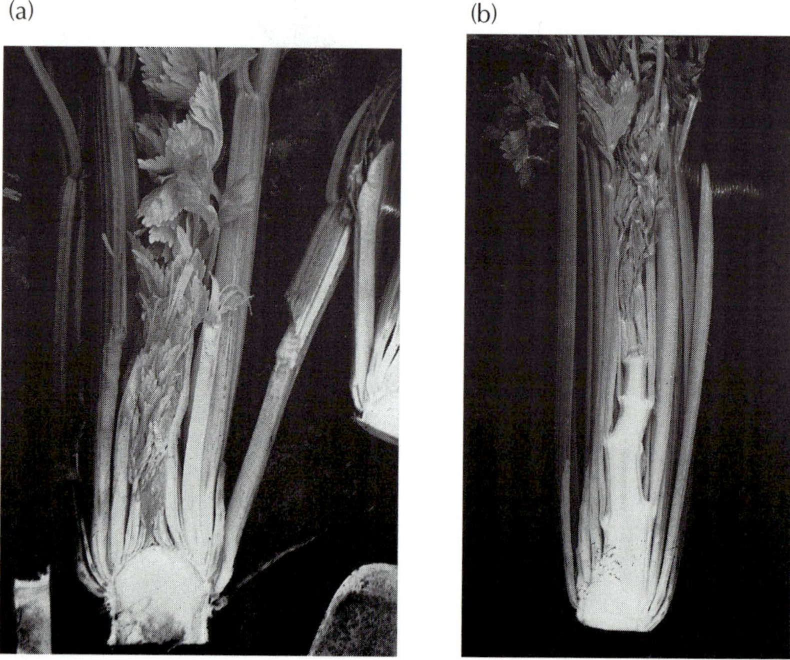

Fig. 2.11. Longitudinal view of celery stem in (a) vegetative state and (b) reproductive state.

Smallage

Smallage plants resemble celery, although the leaves are smaller, taproots not as enlarged or as fleshy as those of celery and the fibrous root system is also smaller. Petioles are generally shorter, less fleshy and often hollow. Leaflet blades usually are smaller than those of celery. The number of leaves approximates those of celery. Smallage foliage is usually dark green, but some cultivars contain anthocyanin. Other plant characteristics are similar to celery and celeriac, although the growth period of smallage is considerably shorter than that for either celery or celeriac.

Parsley

Parsley is a biennial plant frequently grown as an annual. However, when flowering is suppressed parsley can also be cultivated as a short-lived perennial. That cultivation practice is used in order to achieve multiple harvests of the foliage.

Plants typically produce a slightly thickened aromatic taproot and a radical rosette of leaves in the first year that usually grows to heights between 25 and 50 cm. However, the inflorescence of flowering plants easily reaches a height of 100 cm.

Leaves are dark green and shiny, the lowest being long petioled and two to three pinnate with ovate-cuneate or broadly linear-lanceolate toothed leaflets. The topmost leaves are ternate. Several leaf forms produced are identified as plain, curled and fern-leafed, and with some that are intermediates of these forms. The curled leaf forms range from those with lamina slightly to intensely curled and deeply divided, and sometimes are given botanical variety status as *P. crispum* var. *crispum*. Although edible, these forms mostly have garnish and ornamental use. The flat leaf forms referred to as *P. crispum* var. *neapolitanum* have better cold-temperature tolerance and are considered more flavoursome. The fern-leaf forms are not curled but the lamina are divided into many small segments.

At the beginning of reproductive growth the stem elongates, branches and produces terminal compound umbels bearing many small yellow-green flowers. Floral stems are green and solid. Flowers are self-fertile, but cross-pollination usually occurs. Seed appearance and other characteristics are similar to those for carrots and celery, although seeds are slightly curved like those of caraway. About 650 seeds weigh 1 g.

Cilantro

Cilantro is one of several common names for coriander when grown for its foliage rather than its fruit crop. Other names are Chinese or Mexican parsley.

The plant is an annual usually reaching 30–50 cm in height when vegetative, but about 1 m in height when the plant is in the reproductive state. Leaf colour varies from dark green to yellowish green with leaf under-surfaces having a shiny waxy appearance. The rosette foliage of young plants is easily mistaken for parsley. Basal leaves are ovate, although somewhat lobed to tri-pinnatifid with deeply cut segments. Leaves from following nodes are pinnatifid to a higher degree, and the higher the leaves are inserted, the more pinnate they are. The shorter petioled upper leaves are deeply incised into narrow lanceolate or feather-like segments (Fig. 2.3d).

To some people, cilantro foliage has a somewhat objectionable odour which some suggest resembles that of well-worn old shoes. Aldehydric compounds contained in green plants and immature fruit tissues are responsible for this smell. On the other hand, during fruit ripening the aromatic properties change as the aldehydric compounds disappear, and mature fruit have a pleasant orange-like aroma. The flavour of leaves and seed are very different. Seed have a lemon-sage taste, leaves have a tangy or spicy parsley-like taste. Taproots are pale brown, long and thin, and are used as a vegetable in China.

The erect elongated floral stem is nearly cylindrical, relatively thin, hollow and well branched. Stem colour usually changes from green to reddish purple. Leaves also sometimes develop red or purple coloration during the flowering period. As plants age or become reproductive the foliage becomes bitter. The compound inflorescence is short and small with three to ten primary rays of unequal lengths that form a flat umbel. Umbellets may contain 10–20 flowers. Flowers are small with white, pale pink or lilac petals of unequal size (Fig. 2.12a). The peripheral flowers of each umbellet are first to flower and are protandrous, thus ripening is progressive. Central flowers are staminiferous or sometimes sterile. Stigma receptivity can persist as long as 5 days. Cross-pollination by insect activity is common. Self-pollination does occur and inbreeding depression is not observed.

The ribbed wavy-surfaced globular fruit of one type of coriander *C. sativium* var. *vulgare* Alef., grown primarily for seed oil, are about 3–6 cm in diameter. Cultivars of another varietial type *C. sativium* var. *microcarpum* DC., are grown primarily for foliage and commonly produce smaller fruit between 1 and 3 mm in diameter (Fig. 2.12b). The seed oil crops are better suited to slightly warmer climates than the vegetable-type crops. At first the fruit are yellow green and after ripening become light yellow brown. Fruit shattering is common although the fruit does not readily split into two mericarps.

Florence fennel

Although cultivated as an annual or biennial, these perennial plants reach a height of 75–90 cm during vegetative growth. All plant parts are aromatic.

Fig. 2.12. Coriander (a) flowers and (b) fruit.

The fleshy taproot is large, long and well-branched. Stems are highly compressed before bolting. However, from multiple growing points, lateral stems (suckers) are frequently formed. Leaves are alternate with the bright green smooth petioles. Lateral pinnae are usually sessile with feathery leaflets divided into light green narrow pointed lobes. The succulent petiole bases are broad, tightly shingled and clasping (Fig. 2.13). Florence fennel, widely known as finocchio and erroneously as sweet anise or anise, is grown for the bulb-like enlargement of the fleshy overlapping petiole leaf bases. This compact bulbous enlargement usually ranges from 10 to 15 cm in diameter and varies in shape from strongly ovoid to round; the round form is preferred. Tall seed stalks bear numerous small yellow flowers on large open umbels. The seeds are greenish brown, ridged and 5–6 mm long; about 200 seeds weigh 1 g.

Fig. 2.13. Primary edible portion of fennel petioles exhibiting tight basal clasping.

Chervil

Chervil plants are relatively short hardy annuals that grow between 20 and 50 cm in height. Taproots are thin and white. The light green finely compound stem-sheathing leaves have silky lower surfaces. They are three pinnate, the segments are ovate, and each leaflet is divided twice with deeply cut ovate segments, about 10 cm in length. With age the foliage may become reddish. Curly and flat leaf forms are cultivated.

High temperatures promote bolting. Floral stems are upright, branched, hollow, finely grooved and slightly hairy. Umbels of the compound inflorescence are almost sessile at stem nodes with umbellets bearing tiny white flowers. The smooth black fruit contain 5–7 mm long seeds that are thinly ovoid to mostly linear, each with a beak about one-third of the seed's length. About 450 seeds weigh 1 g. Frequent harvesting of the foliage is performed to encourage new leaf growth and to delay the onset of flowering.

Culantro

These plants are erect perennials that grow up to 90 cm in height when flowering. A radical rosette of leaves with spiny toothed leaves 5–20 cm in length is produced during early vegetative growth. The oblong to linear-lanceolate basal leaves have a narrow base, a rounded apex, and are sharply serrate. The foliage, used much like cilantro, has a stronger aromatic aroma than coriander. During the reproductive cycle the stem elongates and at the top divides into three to five branches. The well-branched inflorescence has long stalked umbels with bracts. The minute flowers are greenish white, and the fruit are globose-ovoid and small.

Japanese hornwort

The erect, glabrous, slender and succulent perennial plants (usually grown as an annual) of Japanese hornwort grow from 20 to more than 100 cm in height. Early growth produces a radical rosette of trifoliate leaves that arises from and along the creeping rhizomes. The earliest leaves are largest with very long petioles with sheathing bases. Subsequent leaves have shorter petioles. High plant density production tends to increase petiole length up to 30 cm. Leaflets are sessile, unequal in size, and may appear broadly ovate, obovate and oblique rhomboid, with serrate to dentate margins. The foliage has a pungent parsley-like odour. The root system develops as a creeping clustered growth of long cord-like fibrous roots from a thick rootstock. Small inconspicuous white or greenish white flowers are borne on the many-flowered, terminal inflorescence. The ribbed fruit are narrow ellipsoid in shape and between 4 and 6 mm long.

Water dropwort

Water dropwort is a semi-aquatic and aquatic perennial with erect to ascending growth usually between 25 and 40 cm in the vegetative state, and up to 100 cm or more when reproductive. Both clustered tuberous and fibrous roots are produced. The slender stems are cylindrical, slightly tapering, hollow and sometimes red-tinged and much branched with many arising tiller-like at the base of the plant. Stolons are also formed from which many radical leaves arise. The long petioled alternate leaves are pinnate or bipinnate, glabrous and sheathing; the petioles are hollow. Leaf blades are ovate-oblong, pinnate to tripinnate, and leaflets ovate to narrowly oblong with serrate to entire margins. Upper leaf surfaces are dark green, under surfaces are lighter coloured. Many small white flowers are borne on the terminal compound inflorescence. The ribbed fruit are glabrous, oblong and usually 2–3 mm long and about 1 mm wide.

Asiatic pennywort

Also a perennial, Asiatic pennywort plants produce slender stolons, typically with long internodes. Clusters of ascending petiolate leaves develop at each node, as do fibrous roots. Leaf sizes vary between the two cultivated forms grown. One is a small leaf, runner-like plant, the other is a larger leaf, bush-like plant. Leaf margins may be smooth, slightly lobed or notched. Inflorescences continue to form throughout growth and consist of one to four simple umbels at each node, with one to several flowers per umbel. Petals are white or rose tinged. The 3–5 mm fruit are slightly broader than long.

Lawn pennywort

Lawn pennywort plants are perennial, mostly prostrate, although some are sub-erect up to 50 cm in height with slender stolon-like stems that readily root at the nodes. Leaves are alternate with leaf blades nearly circular, and are borne upon erect non-sheathing petioles about 6 cm long (Fig. 2.3f). The small leaves, 10–25 mm wide are three- to five-lobed and hairy. Variability in leaf shape and pubescence exist among the many intermediate forms of these plants. Fibrous roots readily form at nodes on the stolons. The umbels are solitary and develop opposite the leaves on peduncles shorter than the leaf petioles. Sub-sessile flowers are greenish white. Yellow to brown seed are produced in the laterally compressed fruit. Seed are glabrous or with short stiff hairs and are slightly larger than 1 mm.

Lovage

The lovage plant is a perennial that grows as tall as 150 cm before it annually dies back. Leaves are dark green, glossy and ternately compounded, each leaflet is wedge-shaped and coarsely toothed near the apex, but entire near the base (Fig. 2.3e). Lower leaves can achieve 60 cm in length and a width of 45 cm, but become smaller toward the top of the plant. The foliage has a celery or parsley fragrance. Stems and roots also are aromatic and their extractable oils have a variety of flavouring uses.

Stout erect stems that branch mostly at the terminus are round, ridged, hollow and green or purple. The brown taproots are branching, thick, fleshy and strongly tapered and ridged. The interior flesh is white and pungent. The crown (hypocotyl–root axis) can expand considerably, as much as 30 cm in width after several growing seasons. The inflorescence consists of compound umbels, 5–8 cm wide with nearly equal rays. Flowers are tiny and greenish yellow. The oblong fruit are strongly ribbed. Seeds, about 5–7 mm in length, are yellow brown, flat, oblong-oval, slightly winged and ridged. Plants are usually propagated by division because of the poor germination characteristics of seeds. An unrelated species, *Ligusticum scoticum* known as Scotch lovage shares similar characteristics with lovage.

Black lovage

Also known as Alexanders or horse parsley, black lovage is a biennial plant. During the vegetative state plants are about 50 cm in height, and when reproductive, the floral stems are as tall as 120 cm. The hollow stems bear ovate dark green leaves of three unequally divided segments with serrated margins. Basal leaves with stem-sheathing short petioles can achieve a length of 40 cm. The umbel consist of many umbellets with many greenish yellow flowers. The ovate fruit are small, about 5 mm in length, and black when mature. The edible young shoots and leafstalks used as potherbs have a pungent celery-like taste. Stems are often blanched during cultivation to improve succulence and edibility.

Angelica

Angelica, known as garden or European angelica, is a hardy short-lived perennial often cultivated as a biennial. The plants are tall and range in height from 90 to 200 cm. All plant portions are aromatic with a musky or juniper-like odour, and have a tart, sweet, liquorice-like taste.

Angelica foliage consists of bright green, wide compound leaves with biternate or pinnate divisions; the leaflets are serrate. The large petioles have flared bases that sheath the stem. Basal leaves are sometimes as long as 50–60 cm. The terminal leaflet is three-lobed. The much branched stout stems are hollow and may be green or purplish. The taproot is thick and fleshy. Roots are usually harvested in the fall, whereas leaves and stalks are usually harvested in the spring of the second year. Plants often flower in the second year, and after flowering, plants die. However, selective pruning of floral stems to prevent flowering allows root enlargement to continue.

Globe-shaped terminal compound umbels bear small yellowish green to greenish white flowers. The seed of the ovate-shaped fruit are pale yellow, 5–7 mm in length, noticeably ribbed on one side, and often slightly winged. Seed viability is rapidly lost with storage, and the typically poor germination is frequently due to very slow and uneven seed maturation. Accordingly, propagation is common performed by use of root divisions. **American** or **purple angelica** resembles garden/European angelica in appearance and growth habit, and has similar food uses, but is infrequently cultivated.

Dill

Most dill plants are annuals, although some exhibit a biennial growth habit and are cultivated as such. Plants attain heights of 70–100 cm. The erect, hollow, smooth, cylindrical stems are branching, and usually a greyish green colour. The spindle-shaped taproot usually is slender and does not enlarge. The aromatic finely dissected foliage has a blue-green appearance. Leaves are alternate and two to three pinnate with clasping bases that sheath the stem. The lowest are long stalked and slightly broader than the upper thread-like leaves. Mature lower leaves can be up to 50 cm long and 25 cm wide. Leaflets are highly divided into filiform segments (Fig. 2.3a). Although dill is also an important condiment seed crop, the foliage, commonly known as dill weed, has more extensive usage than the seed.

The inflorescence is a compound flattened umbel, relatively wide and open, and many rayed with long bractless peduncles. Flowers are small and greenish yellow to pale yellow. Fruit at first are green and become dark grey. They are oval, 4–5 cm in length, ridged, slightly winged, aromatic and have a pungent taste. From 400 to 500 seeds weigh 1 g.

Bishop's weed

Bishop's weed is a low erect perennial that grows to a height of 30–40 cm. Plants have stout spreading roots, a hollow furrowed stem with large ovate biternate leaves that are sharply serrated. The flattened umbels bear white flowers.

Samphire

Samphire is a low creeping perennial plant that grows between 30 and 50 cm in height having short thick, ridged stems and fibrous roots. Lower portions of the stem tend to be woody. The glabrous leaves are biternately or triternately compound with short linear, fleshy and glabrous leaflets. Compound umbels bear tiny greenish yellow flowers that produce ribbed ovoid-shaped fruit that are about 6 mm long. Propagation is by division and by seed.

Sea holly

Sea holly is a short (30 cm) perennial with fleshy stiff glaucous bluish green leaves that are broadly ovate, three-lobed with coarse and spiny toothed margins. Flowers are pale blue.

SEED CROPS

Coriander

Coriander is a major condiment seed crop widely utilized for many flavouring uses. Plant characteristics for coriander are the same as those described in the preceding foliage crop section for cilantro.

Caraway

Caraway plants are mostly biennial, but many cultivars exhibit annual growth behaviour. Erect to widely branching plants usually attain a height of about 90 cm. The slender glabrous branching stems are hollow and grooved. Feathery carrot-like foliage is pinnate or bipinnate decompound into narrow linear segments (Fig. 2.3c). Tuberous roots are relatively fleshy, parsnip-like, yellow with a white interior, and as much as 20 cm in length. Roots, leaves and stems are edible, although seeds are the primary product. Compound flattened umbels with many ray peduncles of unequal length bear minute white or pinkish white flowers. The aromatic fruit are black and ridged. Seeds about 5 mm long are dark brown, oblong and crescent-shaped, and noticeably ribbed. About 300–350 seeds weigh 1 g.

Fennel

Phenotypically similar, bitter fennel (*Foeniculum vulgare* var. *vulgare*) and sweet fennel or sweet anise (*F. vulgare* var. *dulce*) are grown for the flavouring seed. These plants are tall perennials, often attaining a height greater than 1.5 m. However, these plants most often are cultivated for seed production as annuals or biennials.

Leaves of their aromatic airy (filigree) foliage are three to four pinnate dividing into linear thread-like end segments. The foliage can be, but generally is not, consumed. Leaves are alternate with long, sheathing, petiole bases that continue almost to the top of the petiole, but which are not wide or fleshy as those of Florence fennel. The cylindrical glaucous bluish green jointed floral stem is solid. Taproots are creamy white and tapering, some with lengths to 20 cm. The umbels are large and wide with long but relatively few rays bearing small bright yellow flowers. Mature oblong fruit with prominent ridges are greyish green and develop relatively uniformly within umbels. Seeds are larger than those of Florence fennel, being 6–10 mm long, light brown, oval, flat and ridged. Another plant known as common giant fennel (*Ferula communis*) is very tall (2–4 m) and produces seeds that are used much like those of bitter and sweet fennel.

Anise

Anise plants are annuals from 40 to 75 cm in height. Their thin, erect, round, branching stems are hollow, grooved and with many fine hairs. Taproots are long and deeply penetrating. The lowermost basal leaves are round-cordate and long-petioled, appearing much like those of flat-leafed parsley. Intermediate leaves are pinnate or ternately compound and toothed, and the short-petioled uppermost leaves are incised into narrow lobes. Overall, the light green foliage has a feathery lace-like appearance. Small, yellowish white flowers are produced in long ray umbellets within long-stalked large loose compound umbels. The grey-brown seeds have a flat ovoid shape, are ribbed, have surface hairs, and are about 3–5 mm long and 1–2 mm wide.

Dill

See preceding discussion of dill in the foliage crop section.

Indian dill

Indian dill is an erect perennial, between 60 and 120 cm in height with downy-white branching stems and very finely divided leaves, although leaflet segments are less fine than those of dill. Petiole sheathing is longer than those of dill, and umbels smaller with fewer rays. Flowers are yellow, producing small ribbed fruit closely resembling those of dill.

Cumin

Cumin is a short annual plant with slender branching erect stems growing to a height of 30–40 cm. The plants are often grown as a winter crop in areas having mild temperatures and low humidity. After flowering, low humidity is very important for seed production and harvesting. The bluish dark green thread-like bipinnate leaves have narrow and sharply tapered linear segments 8–10 mm in length. Petioles are short and sheathing at their base. The long-stalked compound inflorescence produces 5–10 ray umbels with conspicuous bracts. Flower colours range from white to red, but light pink is most common. Dark green fruit are ovoid-oblong, hairy and about 8 mm long and 2–3 mm in width, and strongly ribbed. When mature the fruit becomes yellow-brown. The aromatic yellow-brown seed are narrow and straight with short bristles that easily break away. Several types are cultivated that differ in seed colour, essential oil content and flavour.

Black cumin

Infrequently cultivated, black cumin is found in Iran and elsewhere in the Middle East. Its seeds are smaller and sweeter than cumin and have similar seasoning uses. *Nigella sativa*, which is not an umbellifer, is sometimes called black cumin, black caraway or Roman coriander. *N. sativa*, which is native to the Mediterranean region, has cumin-like black seed that are also used in seasoning.

Sweet cicely

Sweet cicely plants are hardy perennials, usually between 60 and 100 cm in height when vegetative, and as tall as 150 cm when flowering. All plant parts are aromatic and resemble the fragrance of myrrh and anise. Strong, coarsely pubescent green stems are erect, ridged, hollow and branching. Large thick pale brown taproots are fleshy and deeply penetrating with lengths up to 60 cm. The large light green fern-like foliage is thin, soft, silky, with the undersurfaces pale and lightly pubescent. The strongly sheathing leaves are two to three pinnate compound, oblong and toothed, with lanceolate segment lengths up to 30 cm. Umbels with unequal length rays bear many small white flowers. Seeds are shiny dark brown or black, sharply ribbed, narrow and long (20–25 mm). Propagation is commonly by division because of slow and erratic seed germination.

Burnet saxifrage

Burnet saxifrage is a perennial, producing a stout corymb-like branched stem of about 1 m height. Leaves, most being near the base, are pinnate with many nearly sessile oblong to ovate leaflets about 5 cm long that usually are not divided, but are coarsely toothed with pointed lobes. Terminal 10–15 ray umbels are flat-topped and bear white flowers. A form with red flowers is also grown. The oblong fruit are about 2 mm long. The primary use of the seeds is for seasoning; occasionally foliage and roots are used as a potherb.

Ajowan

A winter-grown annual, ajowan can grow to a height of 90 cm. The profusely branched striated stem supports large pinnately divided leaves with clasping leaf bases. The leaf blades are linear acicular segments. The plants begin to flower about 3 months after emergence. The compound umbel contains about 10–15 rays, each with about 10–15 umbellets. The flowers are white and produce aromatic ovoid fruit about 2–3 mm long. Related *T. involucratum* and *T. roxburghianum* plants have similar use, mostly for their flavourful seeds, although the young tender seedlings are also consumed.

TAXONOMY

A number of taxonomic studies of carrot and celery have been reported, and several are continuing. Regrettably similar taxonomic investigations of most other umbellifer species are few and/or limited. Thus, we only discuss the taxonomy of *Daucus* and *Apium*.

Carrot

The contributions of Constance (1971) and Mathias (1971) greatly increased our understanding of *Umbelliferae* systematics. *Daucus* is one of the largest of the *Umbelliferae* genera with about 25 species according to most recent estimates. The majority of *Daucus* species are of Mediterranean, North African or European origin, but a few New World and Australian species are known. *Daucus* has been addressed by several authorities and the relationship within and between wild species and cultivated *Daucus carota* L. have been examined most frequently.

Worldwide resources of *Daucus* germplasm in the wild are extensive, and wild carrot (*D. carota* var. *carota*) is a widespread weedy species in areas of Europe, Asia Minor, Central and South-west Asia, Northern Africa, North and South America. Because many of these locations are far removed from the central Asian centre of origin of carrots, genetic diversity might be limited in some of these regions. However, the use of biochemical and molecular markers indicates there is extensive genetic variation in wild carrot (St Pierre and Bayer, 1991; Vivek and Simon, 1998a, 1998b).

A classification of *Daucus* species with contributions from Heywood (1978, 1983) and Sáenz Laín (1981) is shown in Table 2.1, and their phylogenetic relationships, based upon nuclear random fragment length polymorphism (RFLP) and chloroplast RFLP analysis are presented in Figs 2.14 and 2.15.

Other *Daucus* species are less widespread geographically than wild carrot. Most are Mediterranean, occurring in dry open habitats, near the sea, or in other disturbed areas such as cultivated fields. Their genetic variability has not been evaluated and there have not been any published reports indicating any species as endangered.

Species are identified by differences in fruit shape, size, ridges, appendages and ducts. Pollen shape, bract and leaf characteristics, umbel arrangement and diameter, petal and style size, and chromosome number also assist

Table 2.1. *Daucus* nomenclature, comparative classification, chromosome numbers and distribution of species.

Daucus species	Heywood (1978, 1983)	Sáenz Laín (1981)	Chromosome number	Distribution
aureus Desf.	C	C	22	NM, ME
bicolor Sibeth & Sm. (syn. *broteri*)	PS	PI	20/22	NM, ME
capillifolius Gilli.	D	D	18	SM
carota L.	D	D	18	Worldwide
crinitus Desf.	—	D	22	NM, SM
durieua Lange.	A	A	20/22	NM, ME
glaber (Forsk.) Thell.	D	—		
glaberimus Desf.	D	—		
glochidiatus (Labill) Fisher & C.A. Meyer	A	A	44	AU
gracilis Steinh.	D	D	SM	
guttatus Sibth. & Sm.	D	D	22	NM, ME
halophilus Brotero	D	—	18	SM
hochstetteri A. Braun ex Engles	A	—		NM, ME
involucratus Sibth. & Sm.	—	D	22	ME
jordanicus Post	D	D	20/22	ME
littoralis Sibth. & Sm.	D	—		ME
maximus Desf.	—	PI		NM
montanus Humb. & Bonpl. ex Sprengel	A	A	66	SA
montevidensis Link ex Sprengel (=*pusillus* Michx.) (syn. *hisipidifolius*)	—	D		SA
muricatus L.	PI	PI	22	NM
pusillus Michx. = (*montevidensis* Link ex Sprengel)	L	D	22	NA
sahariensis Murb.	D	D	18	SM
setifolius Desf.	—	M		NM, SM
syrticus Murb.	D	D	18	SM
tenuisectus Coss. ex Batt.	D	D	22	SM

Sections identification: A, *Anisactis* DC; C, *Chrysodaucus* Thell.; D, *Daucus*; L, *Leptodaucus*; M, *Meoides* Lange; PI, *Platyspermum* DC; PS, *Pseudoplatyspermum*. Distribution identification: AU, Australia and New Zealand; ME, Middle East; NA, North America; NM, northern Mediterranean; SA, South America; SM, southern Mediterranean.

celery are largely undetermined. Furthermore, some such as *A. fernandizianum*, may be extinct. Even though celery has a relatively minor status among world vegetables, the need for germplasm collection is evident. A listing of *Apium* species is presented in Table 2.2. A phylogenetic tree based on RFLP markers for the nuclear genome and chloroplast constructed by Huestis (1992) is presented in Fig. 2.16. All the

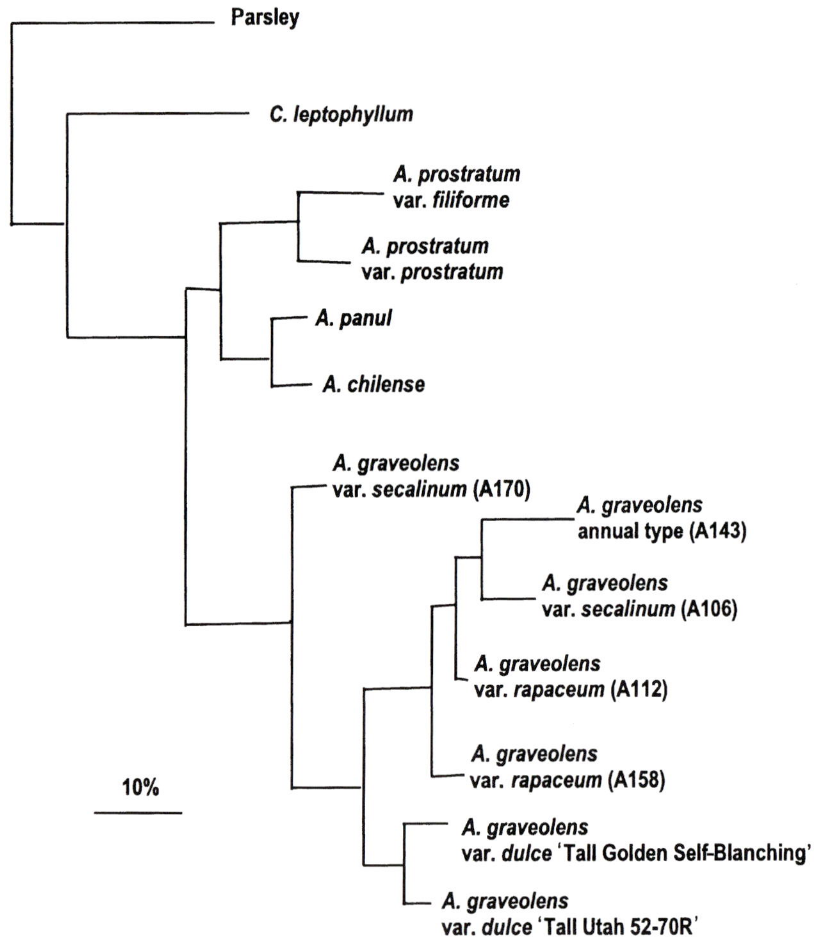

Fig. 2.16. Phylogenic tree of *Apium* species based on RFLP markers constructed with the program PAUP. Outgroups in this tree are parsley and *Cyclospermum leptophyllum*. In parentheses are accession numbers sampled from the University of California-Davis *Apium* germplasm collection. Horizontal branch length is proportional to unit distance. Scale of unit distance percentage is indicated by the bar.

cultivated types cluster together. The wild species are separated into a second main cluster.

Germplasm collections of other *Umbelliferae* species are relatively limited, and when present those specimens are usually maintained in association with *Daucus* and/or *Apium* collections.

3

Plant Breeding and Seed Production

GERMPLASM RESOURCES AND MANAGEMENT

Germplasm availability is an essential resource for carrying out a successful breeding programme for any crop. This resource assures access to a genetic reservoir that can be used by breeders and geneticists to tackle problems presently faced by producers as well as those that will arise in the future. Several umbellifer crops have working germplasm collections that include cultivated as well as wild species, but in many situations these are far from adequate.

Carrot

The extent of carrot and other *Daucus* species held in germplasm collections is relatively small. Approximately 5600 accessions are held worldwide with over 1000 of these at the Vavilov Institute in Russia (VIR) (Frison and Serwinski, 1995). An important *Daucus* germplasm collection is maintained by the Genetic Resources Unit of Horticultural Research International, Wellesbourne, Warwick, UK. Most accessions are *D. carota*, but samples of *D. broteri*, *D. glochidiatus*, *D. gracilis*, *D. hispidifolius*, *D. involcratus*, *D. littoralis*, *D. montividensis* and *D. muricatus* are also held. The US Department of Agriculture collection is maintained at the North Central Regional Plant Introduction Station in Ames, Iowa, where approximately 800 accessions are maintained. Of these, 95% are *D. carota*. Other species held are *D. aureus*, *D. broteri*, *D. capillifolius*, *D. crinitus*, *D. durieua*, *D. glochidiatus*, *D. guttatus*, *D. littoralis*, *D. muricatus* and *D. pusillus*.

Celery

The germplasm available for celery breeding is also limited. A recent International Plant Genetic Resources Institute (IPGRI) report estimates a total of 1270 accessions throughout the world.

Germplasm collections for *Apium graveolens* and other *Apium* species are located in New Delhi, India; Gatersleben, Germany; Bari, Italy; Prague, Czech Republic; and Geneva, New York in the USA (Toll and van Sloten, 1982). The USDA Northeast Regional Plant Introduction Station (NE-9) at Geneva, New York maintains a collection of about 150 accessions, mostly cultivated types and land races of *A. graveolens*. Very few wild species are represented in that germplasm bank. In addition to these collections, the Department of Vegetable Crops at the University of California in Davis, California, maintains one of the largest working inventories of *Apium* germplasm, consisting of approximately 300 accessions. It includes many cultivated types of *A. graveolens* and some wild species accessions, such as *A. chilense*, *A. panul*, *A. prostratum*, *A. annuum* and *A. nodiflorum*.

Other umbellifer crops

Various institutions in the world maintain germplasm collections for other umbellifers. The most prominent are in Braunschweig and Gaterslaben in Germany; Izmir, Turkey; St Petersburg, Russian Federation; and Olomouc, Czech Republic (Frison and Serwinski, 1995).

BREEDING SYSTEMS AND HYBRIDIZATION PROCEDURES

Carrot

Carrot is an out-crossing biennial species but can be handled as an annual in breeding programmes. Carrot has no self-incompatibility system, but inbreeding depression is severe in many wild and domesticated germplasm stocks. Nevertheless, it has been possible to select lines which are able to withstand severe inbreeding and in a few cases beyond ten generations of self-pollination. As these stocks are used as parents in crosses with untested germplasm and with each other, it can be expected that future carrot breeding will include more inbred parental lines with acceptable vigour. An increasing demand for cultivar uniformity, driven by more precise planting equipment and value-added products will no doubt accelerate the development of highly inbred carrot lines.

The discovery of two distinct genic–cytoplasmic types of male sterility (CMS) has provided a system for commercial production of hybrid carrot cultivars. The brown anther type of sterility, first discovered by Welch and Grimball (1947), results in anther degeneration and sterility, whereas the petaloid type of sterility, first discovered by Munger (1953, unpublished report), results in a replacement of anthers by a second whorl of petals (Fig. 3.1).

Celery

The celery genome as described by Murata and Orton (1984) consists of 11 large chromosomes: nine sub-metacentric, one metacentric and one telocentric. The genome size of celery is fairly large compared to carrot, approximately 3.5 pg DNA per 1C nucleus or 3×10^9 bp per haploid nucleus (E. Earle, 1995, personal communication). Cytologically, a single pair of chromosomes in the genome organizes a nucleolus containing the 18S–25S rDNA gene family. It is formed by a tandem repeat unit of 9.3 kbp which is invariable in all three *A. graveolens* cultivated types. There is, however, variation for unit size and restriction sites in the *Apium* wild species tested. A locus for the second unit of transcription for rDNA genes, which is constituted by the 5S RNA gene family, is located on a different chromosome. Variation for these genes is based on differential cytosine methylation in some celery accessions (Yang and Quiros, 1994).

In addition to the 18S–25S and 5S rRNA genes, few repetitive coding sequences have been observed after restriction fragment length polymorphism (RFLP) mapping based on celery cDNA library clones (Huestis *et al.*, 1993). The proportion of single-copy coding sequences in celery has been estimated to range between 59% and 78%. This observation strongly suggests that celery is a true diploid species whose genome has not suffered extensive duplications. The amount of total repetitive DNA in the celery genome has not been determined, but is expected to be similar to or higher than that observed in carrots.

The level of polymorphism in *A. graveolens* is relatively low. cDNA clones tested in celery and celeriac for five restriction enzymes generated only 23% polymorphic loci. A similar level of polymorphism was detected for random amplified polymorphic DNA (RAPD) markers (Yang and Quiros, 1993). The most likely mechanism for RFLP variation in celery seems to be due to base deletion and insertion more than to base substitution.

Other umbellifer crops

Parsnip, parsley and Hamburg parsley cytogenetic research has been limited to chromosome counts ($2n = 2x = 22$ for all) and karyotype description (Kalloo, 1993a, 1993b). Since they are outcrossing species, inbreeding depression is generally severe for parsnip and parsley, but interestingly, not for coriander (Diederichsen, 1996). Cytoplasmic male sterility in parsnip has been utilized for its potential for development of hybrids (Kalloo, 1993a). Relatively little more than chromosome number is known about the cytogenetics of other umbellifer vegetables.

INHERITANCE

Carrot

Carrot is rich in genetic variation among domesticated and wild germplasm. Genetic characterization of variation has been limited by the lack of research effort in this area. Progress has been made in increasing the number of well-characterized, simply inherited genetic loci in carrot. Recent markers include ten isozyme loci, two for seed spines, two for northern root-knot nematode (*Meloidogyne hapla*) resistance and one for purple root colour. This brings the total number of markers to approximately 35 with a large increase anticipated with molecular marker projects underway (Schultz *et al.*, 1994; Vivek, 1997; Vivek *et al.*, 1999; Fig. 3.2).

A number of horticulturally important quantitatively inherited traits in carrot have also been studied that include greening, flavour terpenoids and soluble solids content of the storage root. Genetic variation is known for many other traits, but inheritance has not been reported.

The recognition of distinct male sterile cytoplasms and the advent of

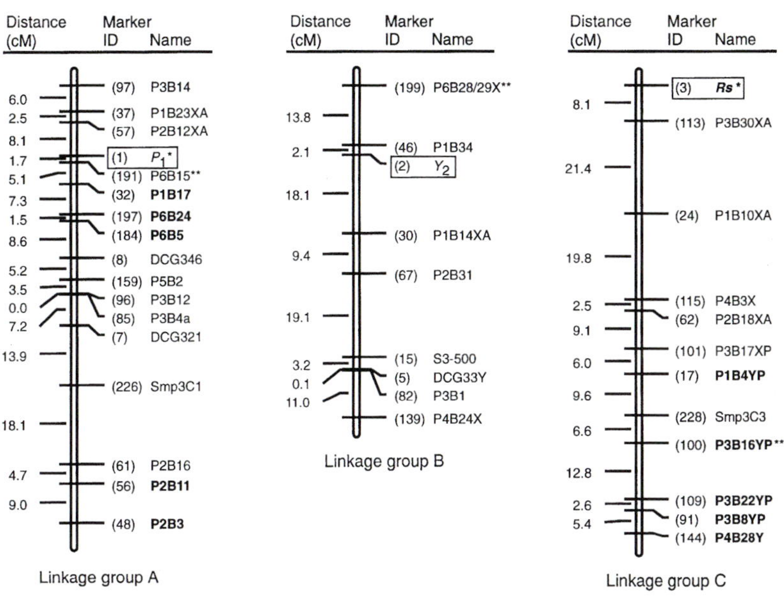

Fig. 3.2. Linkage groups A, B and C of the carrot genetic map. Markers are RFLPs (DCG prefix), RAPDs (S3-500), AFLPs (P prefix) and phenotypic (boxes). Source: Vivek (1997).

molecular techniques has led to several studies evaluating carrot mitochondria and plastid genetic diversity. *Daucus carota* varies substantially with some mitochondrial banding pattern similarities among different CMS types. Plastid inheritance was reported to be paternal, but is now known to be exclusively maternal (Vivek *et al.*, 1999).

Celery

The inheritance of many phenotypic traits has been determined in celery. These include several morphological characters and some disease resistance genes. Such traits have important agronomic and economic value. A description of their inheritance is summarized below.

Hollow petioles

This character segregates as a dominant monogenic trait. Quiros (1993) proposed *Ho* as the symbol for this gene. Hollow petioles are widespread in celeriac and in some smallage accessions. Solid stems and petioles most likely were selected and fixed during the domestication of stalk celery.

Flowering behaviour

Annual habit is a partially dominant monogenic trait determined by the gene *Hb*. This locus forms a linkage group with two isozyme loci and gene *A*, coding for anthocyanin pigmentation in the plant (Quiros *et al.*, 1987). Bouwkamp and Honma (1970) reported that early bolting was dominant over slow bolting, and was determined by a single gene nominated as *Vr*. It is possible that *Hb* and *Vr* are allelic. In contrast to the simple genetic determination of annual versus biennial habit, resistance to bolting in biennial types is a complex trait, most likely polygenic and affected by environment.

Fusarium resistance

Resistance to *Fusarium* yellows (*Fusarium oxysporum* f.sp. *apii* race 2) is partially dominant and determined by two loci (Orton *et al.*, 1984a, 1984b). The resistance based on these two genes has an additive nature, where the partially dominant allele *Fu1* from celeriac contributes the largest effect. The partially dominant allele *Fu2* of lesser contribution to resistance at the second locus is found in tolerant celery cultivars (Quiros, 1987). Based on a series of F_2 and backcross progenies the following genotypes are postulated for disease reaction:

Resistant	Tolerant	Susceptible
Fu1 Fu1 Fu2 Fu2	*Fu1 fu1 fu2 fu2*	*fu1 fu1 Fu2 fu2*
Fu1 Fu1 Fu2 fu2	*fu1 fu1 Fu2 Fu2*	*fu1 fu1 fu2 fu2*
Fu1 Fu1 fu2 fu2		
Fu1 fu1 Fu2 Fu2		
Fu1 fu1 Fu2 fu2		

In order to maximize resistance breeders should aim to fix both loci in the homozygous dominant condition whenever possible.

Male sterility

Until recently, only a single genetic male sterile has been reported in celery. That trait is recessive and determined by a single gene named *ms-1* (Quiros, *et al.*, 1986). This was found as a spontaneous mutant in a weedy Iranian accession. Therefore, substantial breeding activity will be necessary to transfer it into useful genetic backgrounds. Male sterility is due to tapetal degeneration in the anthers. Nectar production in the male sterile mutants is not impaired, therefore the flowers still attract pollinators. Recently, cytoplasmic male-sterility has been reported by Dawson (1993), after finding this trait in unidentified wild celery plants growing in the UK.

Stem colour

Anthocyanin pigmentation is determined by a single dominant gene designated *A* (Quiros *et al.*, 1987). This gene has been found to be tightly linked (approximately 2 cM distant) to the isozyme locus *Aco-1*, coding for the enzyme aconitase. Yellow celery, designated as *y* is recessive to green and it is determined by a single gene (Townsend *et al.*, 1946). This trait often has been used as a marker for hybrid identification.

Leaf shape

Bouwkamp and Honma (1970) reported that deeply toothed leaf is recessive and determined by the gene *dt*.

Other than basic chromosome numbers and some indications of CMS systems, little is known about the inheritance of traits in other vegetable umbellifers.

INTERSPECIFIC AND INTERGENERIC HYBRIDIZATION

Carrot

One interspecific cross has been reported in *Daucus*, that between carrot and *D. capillifolius*, another nine-chromosome species. This cross yielded useful resistance to carrot fly. Other crosses between *Daucus* species have not been confirmed. Carrot protoplasm fusion with parsley (n = 11) has also been successful but hybrid regenerants were not analysed.

Celery

Relatively few attempts at interspecific hybridization have been made in celery because of the unavailability of wild species. Ochoa and Quiros (1989) succeeded in hybridizing celery to *A. chilense* and *A. panul*. The hybridization between these three species can be achieved regardless of which species is used as female. *A. chilense* is resistant to celery late blight and is being used to breed celery lines resistant to this disease (Fig. 6.17a).

The celery × *A. panul*, and celery × *A. chilense* F_1 hybrids are very vigorous but display a pollen fertility range from 0 to 20%. The sterility seems to be due to chromosomal structural changes. The sterility of the hybrids precludes obtaining F_2 progeny, although backcross progeny to celery are readily obtained. After two backcrosses, the petiole characteristics of celery are partially recovered. On the other hand, *A. chilense* × *A. panul* F_1 hybrids are fully fertile and do not show any evidence of chromosomal rearrangements. This indicates that the two wild species have the same chromosomal constitution and are closely related.

Quiros (1990) reported successful hybridization between celery and *A. prostratum*, a species which is resistant to leaf miner *Lyriomiza trifolii* (Trumble and Quiros, 1988), beet armyworm *Spodoptera exigua* (Diawara *et al*., 1994; Fig. 6.17b) and Celery Mosaic Virus. This species is being used to develop insect- and virus-resistant celery lines. The F_1 hybrids are resistant to this pest, quite vigorously, but have a pollen fertility of only 25%. There is no evidence of meiotic abnormalities or of abnormal chromosomal behaviour in the hybrids. *Apium nodiflorum*, a species resistant to *Septoria* and leaf miners has been hybridized to celery by Pink *et al*. (1983). No information is available on the characteristics of the resulting hybrids.

The possible crossing relationships between celery and three wild species are illustrated below by the connecting lines.

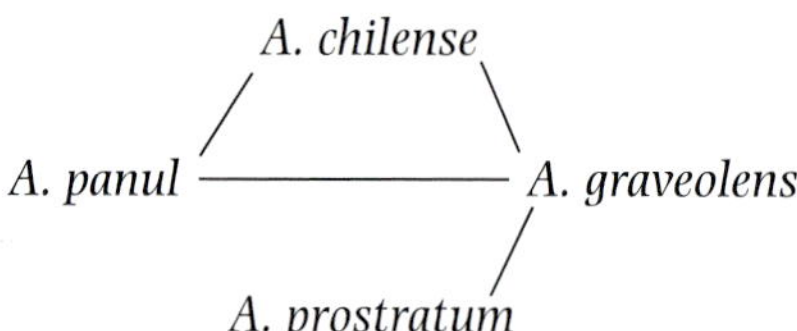

Crossing ability between *A. graveolens* L. var. *dulce* and *A. graveolens* L. var. *rapaceum* was reported by Munger and Newhall (1952). Lamprecht (1961) reported a spontaneous outcross of celeriac and parsley. Madjarova and Bubarova (1978) reported on the successful crossing of parsley cultivar 'Lister' × celery cultivar 'Pioneer' plants to produce a parsley-like cultivar named 'Festival 68'. Reciprocal crosses involving three cultivars of celery and two of parsley produced several forms of leaf celery, each showing considerable type and compositional variation for vitamin C, carotene, essential oil and amino acid content.

Honma and Lacy (1980), with the objective of transferring late blight (*Septoria appicola*) resistance from parsley to celery, crossed celery cultivar 'Golden Spartan' with a parsley introduction (PI 357330). For hybrid detection in the crossing experiment, they used green stem colour from parsley which is dominant over yellow stems normally present in the celery parent as a marker. Three green seedlings indicating hybridization were found among 1000 yellow seedlings germinating from seed collected from the insect cross-pollinated celery parents. However, segregation for resistance occurred in the F_2 populations, and the level of resistance in the hybrid derivatives was weak.

POLLINATION PROCEDURES

Carrot

Because carrot flowers are small, usually bisexual, and bear at most two seeds, controlled pollinations involving emasculation are only rarely attempted. When done, emasculation is performed on centre florets just about to open, and all unemasculated flowers are removed to avoid self-pollination. Umbels of emasculated flowers are then isolated in a cloth pollination cage together with a pollen parent and houseflies, bluebottle fly pupae or newly hatched flies. Individual plant pollination cages are 50–80 cm long wire cylinders, 40–60 cm in diameter, covered with cloth, and open at each end (Fig. 3.3a). They are placed over umbels to be pollinated, and tied tightly with a wire closure around seed stalks at the bottom, just below the umbels. The top is also tied shut. Flies are added through a tube inserted in the top or side. Cross- and self-pollinations are also able to be made in an insect-free area by rubbing umbels of parental stocks together or by moving pollen with brushes or by hand.

To avoid the inefficiency of hand-emasculation, two alternatives exist. If male sterile hybrids are acceptable, use of CMS female parents is an obvious choice. If male fertile hybrids are required, two fertile plants can be inter-pollinated without emasculation. Both hybrid and self-pollinated progeny can usually be found in seed of both parental plants. By planting seed from both plants in adjacent rows, hybrids and self-pollinations can generally be differentiated at harvest by greater root vigour and an intermediate or combination of parental characteristics. This is not always possible if parents are related, similar and/or not inbred, thus leaving hand-emasculation the only alternative. For controlled pollinations involving more than four or five plants, mesh screen cages are used with honeybees or flies added to distribute pollen. Depending on plant numbers, cages sizes of 1×2 m up to 8×30 m are used (Fig. 3.3b).

Celery

Celery flowers are very small, significantly precluding easy removal of individual anthers. Furthermore, different developmental stages of the flowers

(a)

(b)

Fig. 3.3. Pollination isolation cage for (a) single plants and (b) a large number of plants for carrots and other umbel crop plants. Source: Donald Maynard, University of Florida, USA.

in umbels makes it difficult to avoid uncontrolled pollinations. The standard hybridization technique in celery consists of selecting flower buds of the same size and eliminating the older and younger flowers. Then, the umbellets are covered with glycine paper bags for a 5–10 day period, during which the stigmas become receptive. At the time the flowers are receptive, available pollen or umbellets shedding pollen from selected male parents are rubbed on to the stigmas of the female parent. An improvement to this technique consists of washing away pollen and anthers of open, unreceptive flowers with a water stream, covering them with the paper bags and pollinating them 5–6 days later when the stigmas have become receptive. The advantage of this technique is the lower rate of accidental selfing, which is possibly caused by the failure of some anthers to abscise (Ochoa *et al.*, 1986).

For pollen collection, umbels at anthesis are dried at 30–35°C for 14 h, crushed by hand and passed through a sieve (USA No. 100) with openings of 150 nm. Pollen is collected and placed in gelatin capsules for storage (D'Antonio and Quiros, 1987).

Male sterility has been described, but so far little utilized in celery, although cytoplasmic male sterility has been reputed to be used to develop several F_1 hybrid celery cultivars.

An interesting attribute of celery plants that helps to expedite controlled crosses is the ability of cut seed stalks to easily root in containers of water. This allows collection of umbels from selected plants in the field for crosses in the glasshouse or under other isolation conditions. An alternative is to lift selected plants from the field at their vegetative stage, trim back most of the tops and pot them in the glasshouse for continuation of growth. Plant survival is usually fairly high.

Plants require a period of vernalization while in the vegetative phase in order to induce seed stalk development. A period of 6–10 weeks at 5–8°C is usually adequate (Honma, 1959a). However, unless plants are beyond a juvenile state or a minimum of 4 weeks old they may not be receptive to vernalization. Due to a wide range of response to the cold treatment, it is often difficult to synchronize crossing since plants will flower at different times. However, pollen can be stored for 6–8 months at –10°C in the presence of silica gel or calcium chloride with a viability decline of only 20–40%, thus providing flexibility to perform crosses over a longer time.

For selfing, the plant or selected umbels are caged in cloth bags. These are shaken several times during the day to promote pollen release. Houseflies (*Musca domestica*) can also be introduced weekly into the bags to perform pollinations.

Other umbellifer crops

In addition to carrots, some parsnip, celery and parsley hybrid cultivars have been produced. A cytoplasmic male sterility (CMS) system has been reported for parsnips (Kalloo, 1993a) and coriander (Diederichsen, 1996), but they are not widely used. The inadequacy of manageable genetic mechanisms limits attempts at F_1 hybrid production for most other vegetable and condiment umbellifers. Interest regarding interspecific crosses with other umbellifer crops has not manifested itself because of the wide diversity existing among the many different species. Furthermore, the production volume and value of these lesser known vegetables is relatively minor on a world scale. Nevertheless, some interspecific crosses continue to be attempted, with accompanying selection for the improvement of important traits. Advancements in molecular techniques may change the situation.

CULTIVAR DEVELOPMENT

Carrot

The improvement of uniformity and vigour that many carrot hybrids provide is primarily responsible for the gradual replacement of open-pollinated cultivars. The trend is very pronounced in commercial production within Europe and North America, and more so for fresh market production than for carrots grown for processing. Many open-pollinated cultivars are used by growers which do not require the advantages of hybrids or the inability to justify the higher cost of hybrid seed. The increase in market share for seed of hybrid carrot cultivars is expected to continue in the future.

Hybrid carrot seed costs more to produce than open-pollinated seed because a portion (20–40%) of the seed produced in a field is derived from the pollen parent and is not saleable. Additionally, the male-sterile plants used in hybrid seed production often produce less seed per plant than male-fertiles. This is especially true for the petaloid form of CMS. Reduced honeybee attractiveness, due to less nectar or less UV-visible pigment production is thought to account for some of the low seed production of petaloid male-sterile plants. Low pollen production from the inbred male-fertile pollinators and failure of pollen production from the inbred pollen parent to synchronize with the period of seed parent receptivity, 'poor nick', also can be a factor in low hybrid seed production yields. Nevertheless, production from some hybrid cultivars can be as much as, or even exceed, seed production from some open-pollinated cultivars.

Carrot hybrids are usually three-way crosses, $(A \times B) \times C$, since the hybrid vigour in a single-cross F_1 female seed parent usually results in much greater seed production than that of an inbred male-sterile parent (Fig. 3.5). Single-cross hybrids, $A \times B$, are on average more uniform than three-way crosses, and they do not require an extra year to produce F_1 seed parent stock. Thus, if seed productivity of single-crosses is adequate, they are used. To circumvent the reduced uniformity of three-way crosses while maintaining vigour, C.E. Peterson utilized backcrosses as seed parents in several hybrids he released so that the final product was in the form $[(A \times B) \times B] \times C$. This takes an additional year of seed parent production, compared with three-way crosses, but it

	Single-cross	Three-way hybrid	Backcross hybrid
Year 1	A x B	A x B	A x B
Year 2		(A x B) x C	(A x B) x B
Year 3			(A x B^2) x C

Fig. 3.4. Scheme for production of hybrid carrots. A, male-sterile inbred; B and C, male-fertile inbreds.

allows for utilization of less similar seed parent inbreds, A and B, than is usually required.

On a global scale, carrot cultivars fall into two primary categories: temperate and subtropical. The majority of world breeding effort focuses on temperate types because of their larger market share and crop value. Temperate cultivars are marked by lower seedling vigour and slower plant growth rate. They can be grown in subtropical regions but yields often are lower. Mid- or late-season disease effects are more devastating because plant damage usually occurs before plants are well developed. Consequently, any defects in temperate cultivars are magnified when grown in subtropical areas.

When subtropical cultivars are grown in temperate regions, bolting (premature flowering) occurs in a majority of plants so that marketable yield is often poor. This appears to be a response to cool temperatures and/or photoperiod. Thus, subtropical carrot cultivars are grown during the summer in temperate regions and during the winter in warmer climates. In that regard some of these cultivars are utilized for winter production in Florida, Mexico and southern California, as well as for summer production in Canada and the northern USA. In many central California locations these cultivars are grown year-round, although some seasonal preferences are well-identified. A similar level of adaptation is noted for cultivars grown in northern Europe and Scandinavia in the summer, or southern Europe, the Mediterranean and the Middle East during the winter.

Within the temperate and subtropical carrot cultivars there are several classes determined primarily by root shape and to a lesser extent by rate of growth. Temperate types include: 'Nantes', 'Imperator', 'Danvers' and 'Chantenay', while subtropical types include 'Kuroda', 'Brasilia', and 'Tropical Nantes'.

BREEDING FOR SPECIFIC CHARACTERISTICS

Carrot

As a part of the general breeding process, one or more specific characteristics which require more specialized evaluation are usually incorporated for populations under development. Selection for stable maintenance of male sterility and for bolting resistance were already alluded to. Evaluation of both of these traits in climates comparable to production areas is essential for either trait, with progeny testing necessary to ensure stable sterility maintenance. To exercise greater selection pressure for bolting resistance, inbred and hybrid evaluations in more cool winter production areas are often used.

Visual selection of several carrot root and top characteristics is most essential for progress in a carrot breeding programme. This includes attributes of uniformity and appearance listed in Table 3.1.

Carrot breeding goals have historically focused on improving root productivity and appearance as well as seed productivity. More recently consumer quality, disease resistance and speciality uses of carrots are of greater concern for carrot breeders.

Root productivity and appearance

As with every crop, yield is of paramount importance in the improvement of carrot performance. Total yield of carrots is relatively simple to measure but marketable yield evaluation is very difficult due to non-genetic factors such as soil uniformity and condition, plant stand and micro-climatic variation. Genetically influenced variables such as growth rate, root shape and smoothness, exterior and interior colour, crown (shoulder) greening, haulm size and erectness, bolting tendency, and overall uniformity also confound yield and other performance features. The uniformity of carrot hybrids for these variables is a major advantage over open-pollinated cultivars.

A US trend in carrot fresh market production is a greater acceptance of cylindrical 'Imperator' or 'Long Nantes'-shaped roots in preference to the more pointed, conical Imperator type standards which were popular 20 years ago. Specific traits being sought include earlier development of marketable root size and greater uniformity of smooth and dark-coloured roots.

Seed productivity

The uniformity and vigour of hybrid carrot root production was the driving force for the transition from open-pollinated cultivars, even though hybrid seed production was initially much inferior to that of open-pollinated stocks. Low seed yields of the first single-cross hybrids threatened the further development of hybrids. To improve seed yields, three-way crosses have typically been used. Even then, seed yields can be low due to inbreeding depression, low pollen supply and/or short flowering period of some male inbred parents. In

Table 3.1. Important field selection traits for genetic improvement of carrot storage root performance. Emphasis given to these traits differs for different production locations and markets.

Flowering
Low bolting incidence
Tops
Early establishment of true leaves and full, but not excessive canopy
Upright foliage growth
Single dominant stem development, level with root shoulders
Small but strong petiole base attachments
Roots, external
Absence of multiple or large lateral roots (forking)
Absence of longitudinal and other root cracking
Uniform length, shape
Smooth surfaces, small lateral root scars
Minimal presence of small adventitious roots
Absence of green shoulders
Smoothly rounded shoulders, and level with stem base
Avoid sunken or protruding stem
Bright orange colour
Roots, internal
Small core (xylem)
Absence of interior greening (green core)
Absence of cambium yellowing (zoning)
Uniformity of xylem and phloem orange colour
Disease and pest resistance
Alternaria leaf blight
Cercospora leaf blight
Bacterial leaf blight
Cavity spot
Pest resistance
Carrot fly
Leafminers
Aphids
Lepidopterous insects (worms)
Nematodes

addition, some F_1 seed parents are poor seed producers, presumably due to partial female sterility or poor attractiveness to pollinating insects.

Thus breeding for improved seed productivity also is a primary goal for carrot breeding programmes. Petaloid sterility is the CMS of choice for hybrid carrot seed production in the USA since brown anther CMS often breaks down in seed-producing areas with tertiary and sometimes secondary umbels exhibiting some fertility. On the other hand, petaloid CMS has been most often associated with low seed yield, perhaps due to inferior bee attractiveness (Erickson *et al.*, 1979). Brown anther CMS has often been preferred in Europe where sterility breakdown is not reported to be a problem (Bonnet, 1985), and because hybrid seed yields usually are higher than the use of petaloid CMS. Nevertheless, currently used petaloid seed parents produce adequate quantities of seed so that hybrid carrot seed production obviously is feasible, and in fact some preference for petaloid CMS has been reported in Germany (Dame *et al.*, 1989).

Consumer quality

Carrots are not a major staple in any part of the world, but they are commonly accepted as a primary vegetable in many countries. Whether consumers choose to eat carrots is often based on perceptions of quality that include organoleptic, sensory and nutritional factors. Important attributes contributing to organoleptic quality include sweetness, harshness and bitterness, whereas texture as well as organoleptic factors influence sensory qualities. The provitamin A carotenoids are the most important nutrients found in carrots (Simon, 1987, 1990). Cultivar differences have been noted for several attributes of carrot flavour. Knowledge of these differences makes it possible to improve breeding stocks through genetic selection to alter flavour attributes.

Sensory evaluations indicate a large genetic component for variation in raw carrot sweetness. Genetic variation in the content of volatile terpenoids might be used to affect the perception of sugars in raw carrot sweetness as well as harshness (Simon *et al.*, 1980b). Although laboratory methods for volatile terpenoids that contribute to harsh flavour and for sugars affecting sweetness are well-developed, tasting of roots by a trained evaluator is most efficient for screening the large populations produced in a breeding programme. Genetic improvement of carrot flavour has been successful as selection for higher sugar levels and lower levels of harsh terpenoids has been possible without sacrificing other aspects necessary for production (Simon *et al.*, 1980b).

Cultivar differences in bitterness are detectable which also suggests a possibility for control through genetic manipulation. Textural improvement may also be amenable to genetic manipulation (Aubert *et al.*, 1979). Texture can also be readily evaluated by a trained evaluator, but heritability of this trait has not been measured. Genetic manipulation of raw carrot texture is also confounded by the fact that succulent texture desired by consumers appears to impact a high risk of 'root splitting' or 'shatter cracking'. The relationship between texture and cracking is not well understood. As 'peeled' or 'baby' raw carrot pieces become even more popular, long-term postharvest storage may well be expanded. Consequently, opportunities for genetic improvement of storage quality may arise.

Genetic improvement of nutritional value is important since carrots are estimated to provide approximately 30% of the dietary vitamin A in the USA. Improved nutritional value can increase the potential for carrots to be a major vitamin A source in developing areas (Simon, 1992). Major genes conditioning carotene accumulation in carrot roots have been identified (Buishand and Gabelman, 1979), and selection for very high carotene content has been successful (Simon *et al.*, 1989). Current efforts focus on combining high carotene content with superior flavour and developing high-carotene stocks with greater root yield.

An important quality component of carrots for European production is reduced nitrate content. Progress in genetic selection of low-nitrate carrots has been successful in Poland (Zukowska *et al.*, 1997).

Disease and pest resistance

The emphasis on developing sustainable crop production systems with less pesticide use has stimulated breeding efforts for disease and pest resistance. North American carrot breeders have usually emphasized selection for resistance to leaf diseases that include Alternaria blight, Cercospora blight, aster yellows and motley dwarf, whereas European breeders have also emphasized selection for cavity spot, soft rot, carrot fly, nematode, powdery mildew and *Pythium* resistance.

Monogenic resistance to Cercospora blight (Table 3.2) and powdery mildew has been reported, and field evaluation has been expanded for both diseases (Lebeda and Coufal, 1987; Lebeda *et al.*, 1988). Recent research has provided information on the inheritance pattern and selection efficiency for Alternaria blight (Boiteux *et al.*, 1993; Simon and Strandberg, 1998) and for soft rot (Michalik *et al.*, 1992). Researchers at Horticulture Research International (HRI) in the UK have made significant progress in incorporating carrot fly resistance from *D. capillifolius*. A full range of research efforts were combined to identify the best source of resistance, evaluate field resistance over many years, identify a relationship between chlorogenic acid levels and resistance, and develop an integrated pest management system (Ellis *et al.*, 1991; Ellis and Hardman, 1992).

Nematode resistance breeding efforts have also expanded for carrots. Northern nematode (*Meloidogyne hapla*) resistance selection has been underway since the 1970s (Vrain and Baker, 1980) and new sources of resistance have been identified (Frese, 1983; Frese and Weber, 1984). Southern nematode (*M. incognita, M. javanica, M. arenaria*) resistance breeding is also underway, with a good source of resistance in 'Brasilia'; (Huang *et al.*, 1986).

Specialty issues of carrot

In addition to increased per capita consumption of fresh carrots, different processing uses are being evaluated that include juice, dehydration, deep-fried chips and carotene extraction in order to expand carrot consumption. Since calorie production per unit land area per day for carrot is among some of the largest recorded for plants (Munger, 1987), their potential as a biomass source is also receiving some attention (O'Hare *et al.*, 1983). Carrots have also been

Table 3.2. Genes reported in carrot.

Gene symbol*	Character description	Gene source
A	α-Carotene synthesis (may be identical to *Io* or *O*)	'Kinotki' cv.
(*Ce*)*	Cercospora leaf spot resistance	WCR-1-Wisc. inbred
(*Cr*)	Root cracking (dominant to non-cracking)	'Touchon' cv.
Dia-1, -2, -3	Diaphorase isozymes	Misc.
Eh	Downy mildew (*Erysiphe beraclei*) resistance	*D. carota* ssp. *dentatus*
g	Green petiole	'Tendersweet' cv.
gls	Glabrous seed stalk	W-93-Wisc. inbred
Got-1, -2, -3	Glutamate oxaloacetate transaminase isozymes	Misc.
Io	Intense orange xylem	Misc.
L	Lycopene synthesis	'Kintoki' cv.
Ms_1, Ms_2, Ms_3	Maintenance of male sterility	'Tendersweet' cv.
Ms_4, Ms_5	Maintenance of male sterility	'Tendersweet' cv. 'Imperator 58' cv. and PI 169486
O	Orange xylem	Misc.
Pgi-1	Phosphoglucoisomerase isozyme	Misc.
Pgm-1, -3	Phosphogluconate isozyme	Misc.
6-Pgd-2	6-Phosphogluconate dehydrogenase isozymes	Misc.
P	Purple root	PI 173687
(P_1, P_2)	Purple root	Misc.
Rs	Reducing sugar in root	Misc.
(Sf_1, Sf_2)	Spine formation on seeds	'Amkaza' cv.
Skd-1	Shikimate dehydrogenase isozyme	Misc.
Y	Yellow xylem	Misc.
Y_1, Y_2	Differential xylem/phloem carotene levels	Misc.

*Parenthesized loci not named previously; suggested symbol.
Source: revised from Simon (1984).

investigated as a sugar source for this same reason, but the relatively high fraction of non-sucrose sugars and low sugar prices make this development unlikely.

Celery

Celery breeding objectives vary according to the production area, whether green or self-blanching (white) market type and intended uses. Practically all the celery cultivars in production are open-pollinated cultivars. However, a few F_1 hybrid cultivars have entered into production in recent years. Some of these are grown in Australia, especially for processing because of their vigour and large biomass. Overall, the main breeding objectives, regardless of type are uniformity, quality, slow bolting and disease resistance. Table 3.3 lists many of the breeding traits considered in breeding programmes.

In general, breeders attempt to develop cultivars that produce a homogenous population with regard to many important traits. Although the inheritance of traits resulting in acceptable uniformity have not been determined, they most likely are polygenic. Breeders will usually select plants from segregating progenies on the basis of stalk and petiole size, and will fix these traits by subsequent cycles of selfing, and selecting for uniformity in the progenies.

Quality encompasses many traits, namely uniform plants that are symmetrical and cylindrical in shape with thick fleshy petioles of appropriate size. Additional features are absence of pithiness, minimal stringiness, ribbing, susceptibility for petiole cracking and a high level of desirable flavour constituents. The inheritance patterns for most quality traits are largely unknown.

Table 3.3. Important field selection traits for genetic improvement of celery performance. Emphasis given to these traits differs for different production locations and markets.

Flowering
- Low bolting incidence and slow bolting development

Tops
- Tall, upright leaf growth
- Well-shingled and -shaped petioles
- Wide stem base
- Uniform, smooth and thick petioles
- Low presence of petiole ribbiness and/or stringiness
- Absence of petiole cracking and pithiness
- Uniform colour for plant type

Roots
- Deeply penetrating extensive root system
- Self-blanching or green

Disease resistance
- Fusarium yellows
- Late blight
- Early blight
- Western yellow virus

Pest resistance
- Leafminer
- Beet armyworm
- Mites

Furthermore, the environment appears to play an important role in quality expression. For example, Curtis (1938) reported that high soil moisture will increase stringiness in celery due to the formation of stronger collenchyma strands. However, the degree of stringiness varies among different cultivars, which indicates that it has a genetic component. Very little is known about the genetic determination and environmental influence on other traits such as crispness, flavour, shape and ribbiness. Although stem cracking has been attributed to boron deficiency, there are cultivar differences for susceptibility to this disorder.

Two types of pithiness have been recognized. One tends to occur when plants approach maturity, and symptoms appear first in the outer petioles. This type, for which cultivar differences are observed, has also been associated with high temperatures and insufficient soil moisture. The other type of pithiness affects the entire plant and is often evident as hollow petioles. Pithiness is best evaluated by inspecting cross-sections of petioles immediately below the node of leaf blade attachment (Fig. 3.8).

Premature bolting caused by a period of exposure of plants to low temperature results in a rapid deterioration of the petioles which begin senescence and become unmarketable and thus often results in significant crop losses to celery growers. The response to vernalization varies widely in celery. Honma (1959b) reported that the length of time and level of low temperature exposure of the plant are the primary determinants for seed stalk

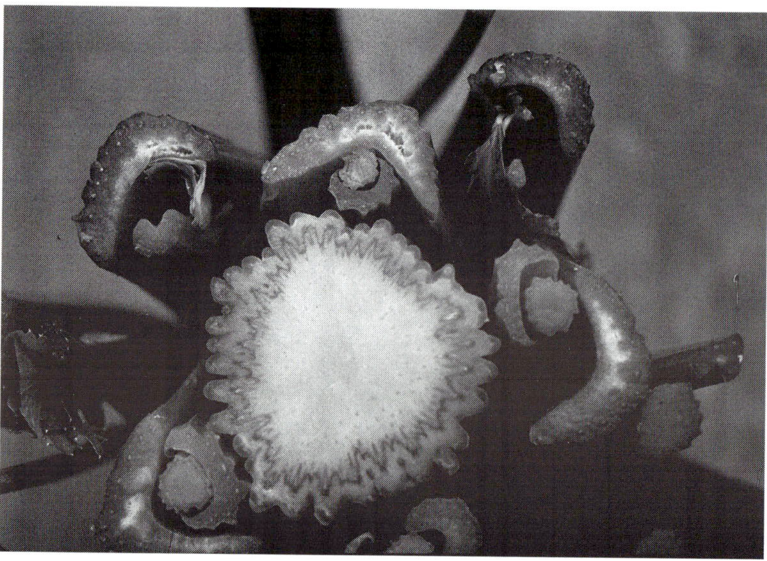

Fig. 3.8. Example of pithiness in cross-section view of celery petioles and stem of elongating seed stalk. Hollow areas and tissue softening result from autolysis.

induction in celery. The effective temperature range usually is between 0 and 10°C, although for some cultivars temperatures up to 14 or 15°C will induce bolting. However, photoperiod length during and after vernalization also influences bolting responses (Pressman and Negbi, 1980). In spite of the difficulty inherent in breeding for bolting resistance, slow bolting lines that withstand low temperature exposure of 6 weeks have been developed by plant breeding and selection (Bouwkamp and Honma, 1970).

Fusarium yellows is an important disease affecting celery in many production areas, and a limiting factor for celery production in California. This soil-borne disease is caused by the pathogen *Fusarium oxysporum* f.s. *apii*, race 2. Although methyl bromide fumigation can be effective to disinfect the soil, it is not an economical or environmentally sound practice. Therefore, the best alternative presently available for growers are resistant cultivars. Most accessions of celeriac are resistant to this vascular disease, and therefore they have been used in the successful development of resistant breeding lines (Orton *et al.*, 1984b) and cultivars. The celery × celeriac cross does not pose any problems of sterility since it is an intra-specific cross. However, the recovery of celery type requires three to four backcrosses.

Late blight caused by *Septoria apiicola* is favoured by cool and moist weather. Although there are cultivar differences in disease reaction, the level of resistance observed is not adequate for breeding. Previous attempts to breed resistance from parsley, and from the wild species *Apium nodiflorum* have not gone further. Ochoa and Quiros (1989) identified two wild species, *A. panul* and *A. chilense*, characterized with a high level of resistance to this disease. The resistance was found to be partially dominant in the F_1 hybrid and segregated in the derived backcrossed progenies. Although the high level of resistance observed in the wild species was not recovered, the resistance of the derived lines is significantly higher than that observed in *A. graveolens*. However, it is unknown whether the level is sufficient to reduce the number of fungicide applications commonly used for control. It is possible that additional genes with quantitative effects may be involved in the resistance.

The development of early blight caused by *Cercospora apii* is favoured by warm and wet conditions. Townsend *et al.* (1946) developed resistant F_2 and F_3 lines from resistant Turkish accessions of celeriac. They found that green celery plants were in general more resistant than self-blanching ones. Resistance seems to be partially dominant and determined by more than one gene.

Little is known about host resistance of aphid-transmitted Western Celery Mosaic Virus disease, although cultivar differences exist. The establishment of celery-free periods intended to disrupt the vectors life cycle effectively control this disease in California and in some other celery production areas.

Other umbellifer crops

Field selection for crop appearance, uniformity, earliness, non-bolting and disease resistance is used for improving other vegetable umbellifers. Disease resistance selection has been successful for black canker resistance in parsnip and for powdery mildew, canker (*Phoma* spp.) and *Septoria* resistance in parsley. Selection for increased oil content of coriander fruit and other vegetable and condiment umbellifers has been successful.

Relative to major world crops, the minor status of vegetable umbellifers is a limitation to concerted research and breeding efforts, with some exceptions in several of the high volume umbellifer condiment seed crops. Information about some of the condiment seed crops often is proprietary and thus not widely or easily available. Other improvement achievements are mainly the results of concerted efforts, observations and experiences from those who cultivate these crops. Overall, most of the minor vegetable umbellifers, although domesticated, receive very little breeding effort for crop improvement.

BIOTECHNOLOGICAL APPLICATIONS

Carrot

Interesting and exciting possibilities exist for the application of biotechnological tools for carrot improvement. As molecular marker systems become more extensive, their use in selecting for difficult or expensive-to-evaluate traits will become useful. Carrot is readily manipulated in tissue culture and genetically transformed, so that opportunities for biotechnologically engineering carrots are also excellent. To date, marker-assisted selection has not been exercised in carrot and transformants have not yet been field tested.

Approximately 1000 RFLPs, 15 RAPDs, 250 amplified fragment length polymorphisms (AFLPs) and 30 other molecular or biochemical (e.g. isozymes) loci have been mapped in several carrot populations to date, with genetic maps from 800 to 1300 cM resulting (Fig. 3.2). Linkages to rootknot nematode resistance, root colour, and root sugars have been established (Schultz *et al.*, 1994; Vivek and Simon, 1999).

Carrot tissue culture systems are very well developed since carrot was an early model system for regeneration and demonstration of totipotency by F.C. Steward in the 1960s. This characteristic is important for facile development of transgenic carrots. Micropropagation is also simple to accomplish for carrot and has some application to increase seedstocks since somatoclonal variation is minimal. However, micropropagation of the carrot crop is not feasible because costs are very high and the incidence of malformed roots through this process is 100%.

Transgenic carrots hold several opportunities, as for other crops, in the transformation of herbicide and insect resistance. Other applications of transgenic technologies and tissue culture include the generation of new types of male sterility, root colour and flavour quality characteristics.

Celery

A number of research areas with great potential are also being developed in celery that include biochemical and molecular markers, tissue culture and transgenics. In addition to the few phenotypic markers described earlier in this chapter, four classes of biochemical/molecular markers are available in celery. These are isozymes, crown proteins, RFLPs and RAPDs (Yang and Quiros, 1994).

Most of these markers have been used to develop a linkage map in an F_2 progeny resulting from crossing celery with celeriac. The present map developed by Yang and Quiros (1994) has 135 loci (33 RFLPs, 128 RAPDs, five isozymes, the *Fusarium* disease-resistant gene *Fu1*, and *Hb* the annual habit gene on 11 major linkage groups (A1–A11) plus nine small linkage groups (A12–A20) (Fig. 3.9). The total coverage of the present map is 803 cM, and the average distance between two markers is 6.4 cM. Two important traits, *Fusarium* resistance *Fu1* and annual habit *Hb*, are located in groups A2 and A7, respectively. The closest marker to *Fu1* is RAPD marker 07-900/800 with a distance of 12.6 cM, which is too large to be used as a tag for *Fusarium* resistance. The existence of a few unlinked markers and more than 11 linkage groups indicates that the current map is still sparsely populated, considering the large genome size of celery.

Isozymes, crown storage proteins and RAPD markers have been useful for cultivar identification, and for following their pedigree and origin (Fig. 3.6). These markers have also been found to be useful for outcrossing determinations (Arus and Orton, 1984) and for hybrid identification after intra- and inter-specific crosses.

Celery is also amenable to tissue and cell culture. It has a high embryogenic capacity and can be regenerated from protoplasts, cells in suspension and tissue explants. An extensive review of this field including optimal media for *in vitro* culture was published by Browers and Orton (1986). These techniques may have application for micropropagation and mass vegetative propagation of parental lines for F_1 hybrid seed production. Another example is the use of somaclonal variation for selection of *Fusarium*-resistant plants (Heath-Pagliuso *et al.*, 1988). Attempts to produce haploids through anther culture have failed in celery and also with carrot. Initial stages of microspore division were obtained by anther culture, resulting in 8- and 16-cell embryoid-like structures. However, these structures failed to develop into plants.

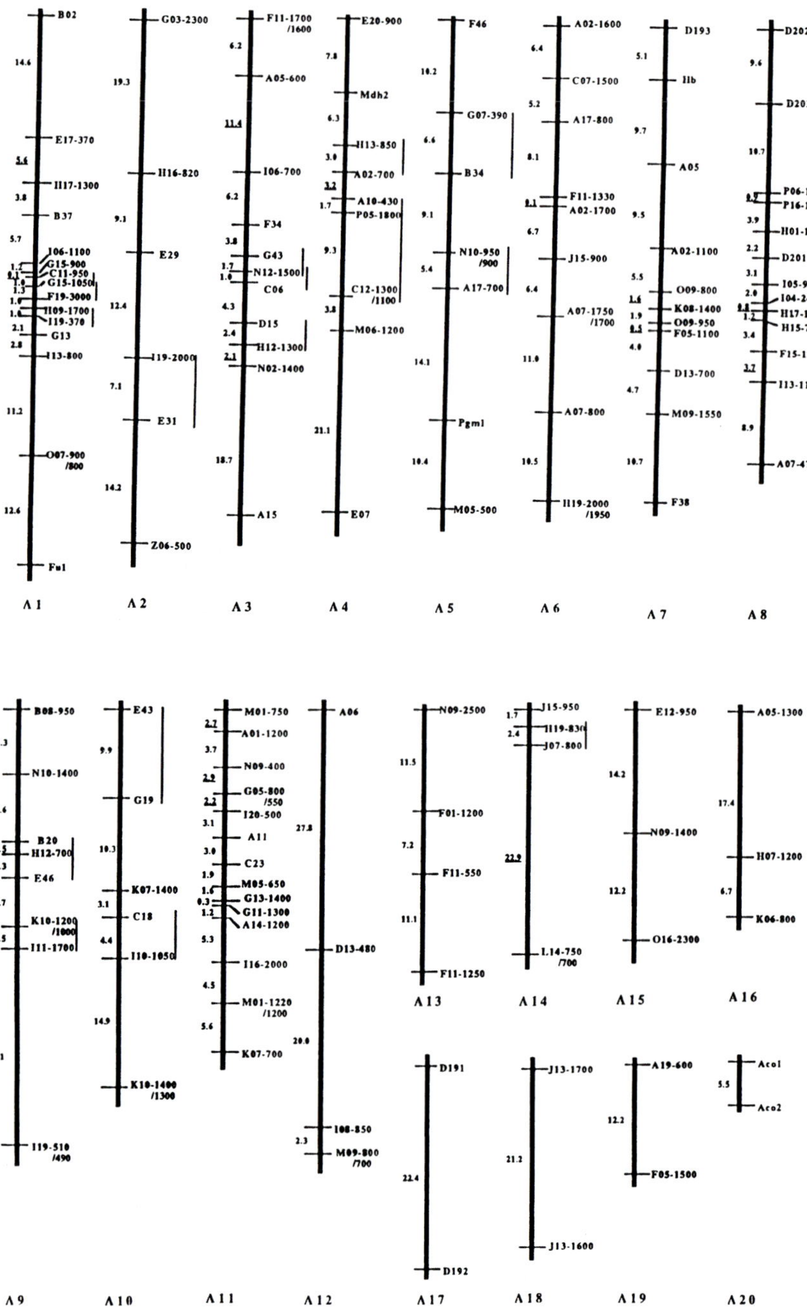

Fig. 3.9. DNA-based linkage map of celery based on RFLP and RAPD markers constructed with the program JOINMAP. The map consists of 11 major groups (A1–A11 and nine minor groups (A12–A20). Vertical bars indicate ambiguous gene order for the specified segments.

Celery has been transformed successfully using *Agrobacterium tumefaciens*, resulting in stable transgenic plants (Catlin *et al.*, 1988). The transgene conferring kanamycin resistance as a selective marker behaved like a monogenic, dominant trait, segregating 3:1 in selfed progenies of the transgenic plants; transgenic celery plants have not been field tested.

These developments in biotechnology place celery at the same level of sophistication as some other horticultural crops. Immediate celery improvement most likely will continue toward the direction of disease resistance, mainly for *Fusarium*, *Septoria*, viruses and insect resistance, especially for leafminers. The availability of a well-developed linkage map will make possible the use of marker-based selection for disease resistance. Similarly, transformation techniques open up the possibility of engineering virus-resistant plants. Other areas of research activity will be the use of wild species as sources of useful traits and their introgression into celery by marker-assisted selection. Development of male sterile lines for F_1 hybrid seed production resulting in the synthesis of dominant alleles for multiple disease-resistant cultivars certainly will continue to be an active research area.

Parsley

Tissue culture preparations, callus production, transformation and somatic embryogenesis are well-developed for this crop. Several genes have been cloned and widely used to study plant disease response mechanisms and gene action (Block *et al.*, 1990).

SEED PRODUCTION

Vegetable umbellifers generally are cross-pollinated, usually by insects, although wind pollination can be a factor for some species. Flowers typically are bisexual, but pollen shed and stigma receptivity are not synchronized and thus do not facilitate self-pollination. Although self-pollination in various amounts can occur, it is usually undesirable except for plant breeding purposes. Pollinating insects are attracted to the flowers and floral nectaries that most species have. An important umbellifer characteristic is the extended period of flowering, where umbels are produced over a long period. For some species umbel formation may proceed for more than 6 or 7 weeks. Additionally, flowering within each umbel is protracted over a period of as much as 1 or 2 weeks. Accordingly, variability in seed development and maturation is prevalent.

The majority of umbellifer vegetables and condiment crops are seed propagated. The variability of seed maturity obviously presents problems for seed producers as well as for growers using the seed and attempting to achieve uniform production of plants whether grown for foliage, storage root or seed crops. Although capable of seed propagation, crops such as arracacha, water dropwort, Japanese hornwort and the pennyworts are commonly vegetatively propagated.

For most large-scale seed production biennial and perennial umbellifer species are grown as a seed-to-seed crop where the crop is sown and remains *in situ* to overwinter with subsequent seed development and harvest following. The root-to-seed method gives breeders an opportunity for evaluation of roots before vernalization, which is desirable and useful for high yield and quality. However, many seed producers forgo some of the quality benefits afforded by the root-to-seed method because of the additional time, effort and expense entailed. Nevertheless, commercial seed production using the root-to-seed method occasionally continues to be employed for carrot, celery, celeriac, parsnip, turnip-rooted parsley and several other crops. Most often the root-to-seed method is used for production and maintenance of stock seed, breeder seed increases, and where winter losses are likely, or for increases of low volume items.

Carrot

Production procedures

The choice of seed production sites is largely determined by the advantages those locations provide. Commercial seed production for carrots and other crops is usually performed where yield and seed quality are optimized. In North America, this is best achieved in the arid Pacific North-west where winter temperatures are low enough to vernalize the seed-to-seed crop, but winter kill is not common and thus plants can be successfully overwintered in the field. When needed, a soil mulching can be provided for freeze protection. In some areas relatively persistent snow cover also provides this benefit. Dry summer weather, adequate irrigation water and a low incidence of wild carrot also recommend this area, which includes eastern Oregon and Washington, southern Idaho and parts of northern California. Significant volumes of carrot seed are also produced for worldwide distribution in southern France, Italy, Israel and Japan on an August-to-August seed-to-seed schedule much like that in the USA. Substantial South American production occurs in Chile. Many other countries also produce seed, mostly as open-pollinated (OP) cultivars for domestic use.

Land preparation for seed-to-seed or root-to-seed production is usually comparable to that for vegetable root production with regard to tillage, soil fumigation, fertilization, weed and pest management. However, overhead irrigation is not applied to carrot plants during seed development in order to keep seed quality high, but adequate furrow or other surface irrigation must be available to minimize moisture stress in order to sustain seed yields.

One special consideration is the need for carrot seed producers to cooperate with other producers in the same area to identify production fields, planting schedules and cultivar types for effective isolation from cross-pollination. This practice was initiated in Idaho by D. Franklin in the 1950s and is widely used by commercial seed producers. Its intention is to achieve 0.5–1 km separation between cultivars of the same root shape class, and up to 2 km between different classes. A minimum distance of 2 km is maintained between fields producing stock seed, although greater distances are preferred. Generally, isolation practices are less critical for other umbellifer crops, although for some like parsley, curly leaf cultivars should be isolated from smooth leaf types.

There is considerable variation about plant populations for carrot seed-to-seed production. Thus, a population range from 20,000 to 50,000 plants per hectare is not uncommon. The logic followed is that fewer plants produce more seed per plant, whereas others believe that fewer seed from more plants will produce a higher total yield. Often the end result is that total yields are similar. It should be noted that in high targeted populations a significant number of plants are lost, and that late germination and slow-growing seedlings produce plants that contribute nothing or little to total yield.

The selection of an appropriate plant density arrangement is very important because of its influence on the distribution of seed yield among umbel orders and potential quality. Some producers use narrow rows and high in-row populations to reduce the number and development of third- and fourth-order umbels. The flowers of these higher order umbels bloom later, are smaller in size, and the seed recovered often have not had sufficient time to mature. A further complication is that flowering is not uniform within an individual umbel. Restricting the development of the third- and fourth-order umbels benefits the seed development in primary and secondary umbels.

Production of only primary umbels results in lower total yield of relatively large seeds. Many studies about the proportion of carrot seeds from different umbel orders show that the major portion of seed in most commercial crops is usually obtained from the secondary umbels. In some cases the harvest of seed from primary umbels may be delayed or ignored and lost to shattering in order to improve the yield of mature seed from the secondary umbels.

Generally, spacing for seed-to-seed plantings ranges from 65 to 90 cm between plant rows and about 3–5 cm apart within the rows. Usual spacing for steckling transplants also are from 65 to 90 cm between rows, but are further apart (10–30 cm) within the rows. Steckling populations are lower because of the greater expense for their production and planting and because stecklings produce larger plants they require more space.

In hybrid seed field production the number and arrangement of pollen parent plant rows relative to seed parent rows varies depending mainly on inbred characteristics and grower practices. A common female to male ratio is 4:1. In the USA this is often grown as an 8:2 arrangement with four two-row beds of female alternating with a single two-row bed of the pollen parent (Fig. 3.10). Field populations of the inbred parents are generally less than the plant population for OP plantings because inbred seeds are most costly to produce and the inbreds fare less well in high density competitive conditions. Inbred stecklings are infrequently used in hybrid seed production because they are expensive to produce, have poor transplant performance and it is more difficult to achieve synchronization of flowering.

Because of the influence of environmental and cultural variables upon seed crop performance it is difficult to predict carrot seed yields. The interaction of plant genotype with environment and cultural variables adds to the unpredictability. However, a reasonable estimate is that most OP fields generally produce 800–900 kg ha^{-1} of seed, and some exceed 1200 kg. Yields from hybrid fields usually are one-third to one-half of those from OP fields, but can range to more than 700 kg ha^{-1}, and occasionally 1 t ha^{-1} is achieved. In hybrid carrot seed production the male fertile (pollen-producing) rows are commonly destroyed before harvest of the female parent plants to reduce possible contamination from sib-seed. Removal of the pollen parents also reduces the possible pollination of late-forming umbels of the female parent. Seeds from these umbels usually will not mature before harvest, and thus will contribute to a reduction of the overall quality of the harvested seed.

Having fewer seed-bearing plants in the field is one of several reasons why hybrid seed yields are often lower than those obtained in OP production.

Fig. 3.10. Field row arrangement for the production of hybrid carrot seed. Note rows of shorter male fertile plants interspersed between rows of taller male sterile plants. Source: Donald Maynard, University of Florida, USA.

Another is inadequate pollen production or 'poor nicking'. Poor nicking is the situation where the flowering of the male 'bull' pollen plants is not well synchronized with flowering of the female plants. Too few or ineffective pollinating insects, and/or cultural and environmental impediments are other reasons for low seed yields.

Commercial carrot seed production of temperate cultivars begins with seed planting in the late summer or early autumn of the first year and proceeds until the summer or autumn of the next year when the seed crop is harvested. In the root-to-seed cycle the seed can be planted at any time of year, although the time of planting will influence the length of cold storage holding of the harvested stecklings. The roots are harvested when beyond the juvenile stage, but before roots achieve full size. If harvested too young, roots are less likely to respond to vernalization or be successfully stored, whereas large roots require more handling and storage space.

In some situations the harvested steckling roots need not be stored because soon after lifting they are replanted. If not already vernalized these roots will achieve that status while in the field after they resume and continue growth. This procedure avoids expensive storage and handling if field conditions are favourable for successful overwintering of the roots and re-establishment of growth.

When field conditions or schedules prevent immediate replanting, the stecklings are held in cold storage. Following harvest and before cold storage, stecklings that pass initial selection are trimmed, with the petioles cut back to within 2 cm of the apical growing point (stem apex). Narrow root tips of long strongly tapered roots are also trimmed because such root tips tend to increase root desiccation, and excessive root length can interfere with replanting. Roots, uniform in size, about 2–4 cm in diameter at the shoulder are preferred. Smaller roots are prone to weather loss, soil heaving, late to bolt and are generally less productive, whereas very large roots often incur crown damage. Accordingly, roots that are over- and under-sized, off-type, damaged and diseased are discarded. The presence of any leaf or root fungal attack before root lifting generally necessitates use of fungicides. Stecklings often are dipped in a fungicide solution, and air-dried completely before cold storage.

The trimmed roots are stored in a low temperature chamber at high relative humidity. The optimum temperature for vernalization of many cultivars is 5 or 6°C. However, lower storage temperatures of 0–1°C are better for maintaining steckling condition. Low temperature exposure for approximately 8 weeks is adequate to vernalize most cultivars. If field conditions or planting schedules require a delay in transplanting, the stecklings can be stored for a longer period. Strict attention to low temperature and high relative humidity conditions is essential to maintain the stored stecklings in good condition. During cold storage, roots are checked periodically to avoid wet or flaccid roots. Following the storage period and removal from storage, further selection may be made to discard damaged and diseased roots (Peterson and Simon, 1986). Additionally, internal colour, core size and flavour selection can be made, and laboratory samples may be taken from roots with little compromise of successful transplant re-establishment as long as at least 8 cm of root length is retained. The steckling method is also used for celeriac and parsley seed production. Clearly, root-to-seed procedures for seed production represent sizeable investments of effort and capital.

Stecklings are taken from storage and transplanted when growing conditions are favourable in the spring so that the resulting seed crop will be harvested in late summer or early autumn. Essentially this is a 2-year generation cycle, since root production from a spring planting requires that the roots be stored after lifting in autumn until spring transplanting to produce the seed crop. However, for some locations production time is possible when sowings for steckling production can be made in late summer or in the autumn. In these locations plants can achieve adequate growth before steckling roots need to be harvested from the field. The advantage of this practice is a reduction in the length of cold storage holding compared with spring sowings. The minimum storage period of low temperature exposure for vernalization is similar whether obtained in the field or in storage. For some situations glasshouses are used for seed production during winter. However, this limits seed production to a small volume, and results in an expensive seed crop about 6 months out of synchrony with most field-grown commercial seed production.

Healthy plants, adequate pollinators and isolation of different stocks are necessary for successful production. Reliance on native insect pollinators is not common or prudent. Both seed yield and viability are positively associated with the effectiveness of insect pollination. Thus, 15–20 beehives per hectare are commonly placed bordering the production field. Although more labour demanding, hives within the field are more effective. Breeders commonly use houseflies, bluebottle flies or bees to pollinate plants that are isolated in cages (Peterson and Simon, 1986).

Seed crop harvesting

Scheduling seed harvest is a major decision that requires considerable planning because the flowering behaviour of umbellifer crops results in the production of a wide range of seed at different stages of development and maturity. The percentage of seed from different umbel orders varies and is influenced by plant population and harvest timing (Table 3.4). Additionally, because seed shattering is a common characteristic of umbellifers, harvest delays intended to increase yields can concurrently increase the amount of shattering of early matured seed. Although total seed yield may benefit from a delayed harvest, it is at the expense of seed quality. Carrot seed crops tend to shatter less than celery or parsley and those of parsnip have a relatively

Table 3.4. Plant density effects on total yield of cleaned carrot seed and the percentage contributed by primary, secondary and tertiary umbels (averaged over three harvest dates each year).

Year	Density (plants m^{-2})	Total yield cleaned seed (kg ha^{-1})	Percentage of seed by weight from		
			Primary umbels	Secondary umbels	Tertiary* umbels
1978	10	1286	18	74	8
	80	1632	63	37	0
		SED = 97			
1979	10	966	38	58	4
	80	1360	57	43	0
		SED = 194			
1980	10	2414	19	74	7
	80	2376	65	35	0
		SED = 308			

*Calculated by subtraction.
Source: Gray *et al.* (1983).

high incidence of shattering.

Determination of the appropriate harvest time varies between producers even when made after sampling to determine crop condition, seed quantity and quality. Agreement about which are the most suitable harvest period indicators is lacking. Most often producers use seed colour development as a maturity indicator: sometimes harvests are made when seed of the primary umbels are brown and the secondary umbels are beginning to turn brown. In other cases, harvest begins when seed of the secondary are brown and seed of the tertiary umbels begin to show a brown colour. Another, generally less precise, indicator as to the appropriate time to harvest are dry and brittle inflorescence stems. Similar criteria are used to determine when seed of other umbellifers are harvested. There are other factors, such as the threat of rain or strong winds, that can influence harvest scheduling. Lodging can be an important concern because seed stalks even when relatively sturdy are broken or flattened by strong winds. High soil fertility or high levels of nitrogen fertilizers and high density populations tend to increase seed stalk height and tall plants are more vulnerable to lodging.

Where production quantity is small or where machines are not available, hand harvest and hand thrashing (flailing) is still practised. The plants are collected from the field and piled upon canvas or a smooth dry surface. This procedure allows for additional drying and seed maturation, and also captures seeds that shatter. Piles should not be too high so as not to interfere with ventilation and drying. The drying period usually is 4–7 days during warm and dry conditions, but may be extended during high humidity or low temperature weather. Rain and wind are threats to the loss of seed and seed quality.

In other situations, the plants are pulled or the inflorescences are cut and swathed by machine or hand into windrows to allow additional field drying and for further development of late-maturing seed. After the drying period the windrows are picked up and thrashed in the field or taken to stationary thrashing machines for seed separation and collection.

Commercial-scale seed harvest of carrot, celery, parsley, parsnip and other umbellifer vegetables is generally accomplished with machines specifically designed for thrashing and separation of dry seeds. Various types of harvest machines are used, many of which are mobile combines that remove the plant or cut off umbels and convey them into a large rotating thrashing cylinder with rubber-covered bars that beat the seed free from the umbels. Fans and screens separate seed from plant trash. The seed are discharged into collection containers and the trash is returned to the field. Other machines utilize the principle of rubbing the umbels to release the seed. Occasionally, to assist harvesting and reduce the biomass handled, the plants may be chemically defoliated before combining.

Most often all the plants in a seed field are harvested at the same time, although occasionally early developed primary umbels are hand-harvested and the remaining umbels permitted to continue development for a later and/or final harvest.

Seeds are cleaned by further separation from umbel parts and other accompanying trash using air flow or other seed-cleaning methods. In most commercial seed-handling facilities, gravity separation equipment is used. Because weight and seed size are well correlated, such equipment when well operated, is very effective in cleaning seed and making good size separations. Precisely sized screens are also used for size separation.

Seeds of carrot, dill and caraway have spines which require removal (milling). This is performed with equipment that gently rubs the seed. If not removed, these appendages cause seed entanglement and clumping, making further separation difficult and also interfering with precision sowing. Seeds of celery, parsley, fennel and several other umbellifers, although ribbed, are not spined and do not require milling. Parsnip and angelica seeds have small wing-like ribs, but these generally do not present a separation problem. Coriander differs from other umbellifers because, except for propagation uses, the marketable product is the intact fruit rather than the separated mericarps.

After cleaning, the seeds are carefully dried to the moisture content optimum for the particular crop. If not used relatively soon, the seeds are stored and maintained at a temperature and relative humidity best suited for extended holding. For carrots, seed moisture ranges from 7% to 10% depending on the temperature and relative humidity conditions of the seed storage. This is fairly representative of most umbellifer seeds. If necessary the seed moisture content can be adjusted by re-hydration. With good storage conditions, seed of some species, notably celery, have relatively long viability whereas others like parsnip are short-lived. Seed should be clean and disease- and pest-free.

Celery

Much like carrot, commercially produced celery seed is also grown in mild winter regions. Most often the seed crop is first established with seedling transplants during the autumn which allows for substantial plant growth before exposure to winter temperatures that will provide sufficient in-field vernalization for seed stalk induction. In the spring, with warming temperatures and a resumption of more rapid growth, the seed stalks elongate, flowers and seed are produced and the seed can be harvested in summer or autumn. Transplants are used to reduce the in-field growing period compared to the longer period of field occupancy required in a field-sown seed-to-seed programme.

An alternative practice of harvesting young celery plants much like that of carrot stecklings for cold storage holding and vernalization during the storage period is seldom performed, except for stock seed production, and for breeding programmes. In this procedure plants are lifted from the field and the foliage is topped, leaving an 8–10 cm length of petiole and an adequate amount of root; usually a minimum of a 5–8 cm length of taproot. The optimum storage-holding conditions for seed stalk induction are temperatures of 6–8°C for 5–6 weeks with high relative humidity. Fluorescent lights to provide a 12 h photoperiod are beneficial. If plants are to be stored for a longer period, temperatures are lowered, and high relative humidity conditions are continued.

Following vernalization, the plants are replanted about 0.8–1 m apart in the second year to allow for the reproductive cycle of growth for seed stalk elongation, flowering and seed development. This procedure is more lengthy because the first year is occupied with the growth, evaluation and selection of the plants that will be grown for seed. Overall, the other field cultural procedures for celery seed production are similar to those used for carrot. Occasionally, for small volume increases, plants are grown in a glasshouse. In addition to low temperature induction, artificial lighting to supply a long-day photoperiod will hasten bolting of the plants.

During field production a wide variety of flies, bees and wasps are attracted to the flowers and they are effective pollinators. Supplementation of insect pollinators using beehives is beneficial. Isolation distances to avoid pollen contamination should be at least 3 km in consideration of the flying range of the insect pollinators. It is necessary to provide pollinators, usually houseflies, when plants are grown in a glasshouse or other confinement facilities.

Seeds attain full size about 5–6 weeks after pollination and become fully mature at 8–10 weeks. Harvest determination and practices parallel those for the carrot seed crop. When working with a limited number of plants, plant breeders often will cut the umbels leaving 30–45 cm of stem. Several stems are bundled together and hung with umbels upside down to allow for drying and late-maturing seeds to develop fully. The bundles are placed in a well-ventilated location protected from light and wind. After an adequate period of drying, the seed is thrashed and cleaned. Seed is brought to a moisture content of about 8–9%, and stored at 4°C and 60% relative humidity. Seed viability tends to be very good, in some instances for as long as 10 years.

Other crops such as celeriac, smallage, parsley and parsnips are handled similarly, with small variations in seed harvesting, handling and storage practices. Harvest of most OP umbellifer seed crops is relatively simple when grown in pure stands and all plants are harvested. On smallholdings, the seed crop occasionally is grown in mixed populations with other vegetables or grains. In these situations, harvesting practices may require some modification. Perennial umbellifer vegetables like arracacha, Japanese hornwort, water dropwort and the pennyworts are seed producers, but they are increased by vegetative propagation practices.

SEED STORAGE

High moisture and high temperature cause rapid deterioration of seed and these conditions become more critical the longer the seed are stored. Drying of many umbellifer seed crops can be accomplished in 1 or several hours, but this depends on initial seed moisture and the volume being dried, as well as dryer capacity. The initial moisture content of seeds is commonly lowered by drying in an air flow at 50°C unless moisture content is greater than 25%. In that situation, the warm drying air temperature should be reduced to 40°C. Storage temperatures after drying need not be very low, 10°C is satisfactory when seed moisture content is low, and the relative humidity of the storage environment is also low. Under such conditions, dried seed of many vegetable umbellifers at the moisture content appropriate for each species usually can be expected to retain their germination and vitality qualities for 3 or more years when placed in sealed moisture-proof packaging. Notable exceptions are parsley and parsnips which lose viability rapidly and generally do not store well beyond 1 year. For extended seed preservation, maintenance at sub-zero temperatures is required.

4

Plant Growth and Development

INTRODUCTION

Sustained plant growth is dependent on the availability of moisture, light, temperature, nutrient minerals and carbon dioxide (CO_2). A deficiency of any one of these factors is growth limiting. Plant shoot and root systems typically maintain a complementary relationship. Photosynthate intended for the fibrous roots and shoot apical meristem are translocated to the storage organ. Likewise, water and mineral nutrients are translocated through the storage organ to the leaves. As the plant leaf area expands, water loss through transpiration also increases. The increase of water loss is replenished by greater water absorption from an enlarging root system. Leaf, shoot and root enlargement also requires greater amounts of mineral nutrients which are absorbed along with the soil water by the root system. A prolonged deficit or excess of soil moisture will limit growth. Shoot and leaf tissues utilize light energy, CO_2 and water in the process of photosynthesis that produces the photosynthates used with mineral nutrients to form new plant tissues. Oxygen is important in the respiration processes of all plant tissues and for root absorption of soil water and minerals.

Temperature, through its effect on physical and biochemical processes, influences the rate of plant growth and development. Most temperate-region crops will grow in the range of 4–40°C, although growth becomes limited at either extreme. Growth processes are slow at low temperatures, and growth rate becomes more rapid as temperatures increase. However, when temperatures are high, net growth is curtailed by an increase in the rate of respiration which if continued may also deplete plant food reserves.

Being mostly temperate cool-season crops, the yield and quality of umbellifer vegetables whether grown for their storage roots, stems, leaves or seeds are diminished when mean temperatures are above 25°C and/or below 10°C. In addition to possible initiation of premature bolting, low temperatures reduce plant size and quality characteristics. Frosts can kill or damage seedlings, mature leaves and exposed root crowns which will impede further growth, and harvest practices. Excessively high temperatures also damage plants and may affect quality by increasing the development of harsh flavours, reducing sweetness and increasing fibrous texture. Excessive heat during flowering will also reduce seed set.

GERMINATION AND EMERGENCE

For most umbellifer species, seed germination and early seedling growth tends to be slow and notably variable within a given seed lot. It is not unusual for seedling emergence to extend over a period of several weeks. This is largely due to the typical staggered flowering behaviour of umbellifer plants that results in variable fruit maturation. Presumably this is an adaptive mechanism to stagger seed germination through a range of climatic and plant competitive conditions. However, growers view this as an undesirable inconvenience because it can reduce uniformity to the point that late-germinating seeds produce non-productive seedlings that often become weed-like competitors to earlier emerged seedlings.

The initial step in germination is water imbibition which begins the biochemical processes of synthesis and respiration in order for the rudimentary embryo to elongate and grow. Other obvious factors influencing germination are appropriate temperature, stage of seed physiological maturity, seed coat permeability, absence of anaerobic conditions, salinity and seed dormancy. Dormancy is the failure to germinate in the presence of normally favourable environment conditions. Light exposure or a period of seed postharvest exposure to low temperature may be required for some species before germination can be initiated. Celery seeds contain inhibitory compounds that prevent germination until they are dissipated or removed. Seed from primary and secondary umbels are reported to be less dormant than those from tertiary and quaternary umbels (Thomas *et al.*, 1978b). Celery cultivars incur thermodormancy which results when high temperatures during imbibition prevent the completion of germination. Even with favourable conditions, seeds of vegetable umbellifers characteristically are intermediate to relatively slow to germinate, and seeds of several species are especially slow. Variation in germination of sown seed initially occurs because of seed-to-seed differences in their rate of germination.

The vigour and percentage of seeds that germinate are important aspects of seed quality. Components contributing to seed vigour include genetic (genotype) constitution, environmental conditions during seed development, and the nutritional status and health of the parent seed plant. The stage of seed maturity at harvest, seed size and composition, mechanical integrity, and possible deterioration due to seed ageing all influence vigour. Successful emergence is the process whereby the radicle emerges and the embryo develops into a self-sustaining seedling. The rate and uniformity of seedling emergence are other primary attributes of seed quality. These qualities are influenced by environmental conditions. Favourable temperature, soil moisture, depth of soil cover and the absence of soil salinity, compaction, crusting or other impediments benefit emergence and early seedling performance. When these factors are favourable a greater level of uniform emergence is more likely to result.

Benjamin (1982) reported that carrot marketable yield was influenced not only by spacing between plants, but also by the time when each becomes established, and seedling size at emergence. Gray (1984) reported that carrot seeds with a high germination percentage (>90%) were associated with shorter mean germination times and a narrow spread of germination time while the opposite was true for seeds with a lower germination percentage. Seed lots with a low percentage of germination generally resulted in greater variability in plant weight than those of a higher percentage of germination. Seeds with low viability exhibited slower emergence, a greater time to emerge, and a greater variation in plant size than those of high viability which could result in significant differences in root size at harvest. Gray (1984) concluded that a proportionate increase in the number of seeds sown to compensate for differences in initial viability will not completely eliminate differences in performance between seed sources.

Salter *et al.* (1981) showed when the spread in the time for carrot seedling emergence is shortened, the early emerging plants have more rapid and more extensive use of nutrient resources and therefore are stronger competitors than late-emerging plants. Thus, when the spread in emergence time is lengthened, crop performance is adversely affected. Variation in carrot seedling size or weight may result from variation in seed weight (mostly endosperm), embryo size, rate of germination and length of germination period. Variation of carrot root weight at harvest was related to variation in seedling weight soon after emergence.

SEED QUALITY

In addition to genetic constitution, many events during the seed plant's growth will influence resulting seed quality. For seeds of carrot and other umbellifers, the umbel position on the mother plant and time of harvest have a strong influence on seed development and performance. Seeds develop sequentially through endosperm and embryo formation, expansion, reserve deposition and dehydration. Seed quality also develops sequentially, with germination capacity developing early, followed by desiccation tolerance, vigour and storage life. The first-formed seeds of the primary umbels tend to be physiologically more mature, and usually larger and heavier than those that develop later (Gray and Steckel, 1980). The primary umbel seeds usually have a larger endosperm so that the ratio of endosperm to embryo is larger. When harvested at the same time, seeds from primary umbels were reported to have a higher germination percentage than similar size seed obtained from secondary and tertiary umbels (Jacobsohn and Globerson, 1980).

The variation in embryo length was lower in seeds from primary than secondary umbels, although there was less variation with later harvested seed, the reduction being greater for seed from secondary umbels. Fewer seedlings emerged, and emergence was later from early harvested and secondary umbel seeds than from late harvested and primary umbel seeds. The late harvested and primary umbel seeds also produced larger seedlings than early harvested or secondary umbel seeds, even allowing for differences in seed weight. The variation in seedling weight decreased with a delay in harvest and was lower for seed from primary than from secondary umbels, especially at the early harvest (Gray and Steckel, 1983a).

Indeterminate flowering patterns complicate production of high-quality seed, as seeds at different developmental stages are present at the same time, and shattering often results in loss of mature seed. Differences in seed performance due to umbel position may be minimized if seed harvest is delayed, which essentially allows seed on later-formed umbels more time for development. However, harvest delay increases the exposure of earlier-formed seed to moisture stress or other adversities that may reduce mature seed vigour, as well as increasing the possibility of seed loss due to shattering.

Heavier, usually larger seeds do not necessarily denote better quality. Gray and Steckel (1983a) showed that variation in carrot seedling weight at emergence was more closely correlated with the spread of emergence and embryo length rather than with seed weight. Embryo size at seed harvest and the variation in embryo size among seeds was directly correlated with variation in carrot seedling sizes and with variation in root size at harvest. Variability in seedling weight and the spread of seedling emergence times were closely related to the variation in embryo length, but not to the coefficient of variation (CV) of seed weight. Thus, management practices to enhance uniformity of embryo size within a seed lot can directly result in higher yields of marketable carrot roots. Variation in seed weight was not affected by harvest date (Gray and Steckel, 1983b).

Gray and Steckel (1983c) also showed that delaying harvest resulted in longer embryos and a reduction in the CV of embryo length, in the spread of emergence time and in seedling weight. Their results indicated that embryo growth continued after the endosperm and pericarp have achieved their maximum weight. Because variation in seed weight and embryo length are less closely related, reliance on seed weight separation to identify quality differences by seed grading is of little benefit.

Final seed weight depends on genotype, volume of the ovary at the time of fertilization, and the rate and duration of the cell division period during seed development. Gray *et al.* (1986) demonstrated that carrot embryo length and cell number exhibit a positive linear relationship which was unaffected by seed crop density, harvest date and seed umbel position, but was affected by the season of production. The seasonal effect possibly was due to temperature differences between years during the seed development period. Differences in rate of germination and percentage of germination were reported to be strongly related to embryo length.

CARROT SEED DEVELOPMENT

During carrot seed development the endosperm becomes cellular at about the time the embryo is at the two-cell stage (Borthwick *et al.*, 1931). Gray *et al.* (1984) reported that a period of rapid cell division of the endosperm along with cell expansion follows after anthesis. Maximum carrot seed dry weight and maximum endosperm volume was reached at about 35 days after anthesis. At this time the endosperm was still soft, the pericarp green, and less than half of the seeds were viable. Fully physiologically mature seed were not produced until about 44 days after anthesis. Seventy per cent of the increase in endosperm volume was due to an increase in cell number which ceased 35 days after anthesis. The increase in embryo volume was slower and was due to an increase in both cell number and cell volume which continues until about 49 days after anthesis. The embryo at maturity was the equivalent of 2–3% of endosperm volume.

ENHANCEMENT OF GERMINATION, EMERGENCE AND SEEDLING DEVELOPMENT

In addition to the advantages of inherently good seed quality characteristics, seed-handling practices and technologies such as seed conditioning and growth regulator treatments are used to enhance the rate and percentage of germination. These treatments also tend to increase the uniformity of emergence.

Umbellifer seeds generally do not require scarification to facilitate imbibition. However, an old seed-conditioning procedure is the pre-germination (chitting) of seed before sowing (Fig. 4.1). In this practice seeds are first permitted to imbibe water while held at a low temperature in order to retard radicle emergence. The treated seeds are usually sown before radicle protrusion is imminent or evident in order to avoid injury.

Seed pre-germination practices are used for other umbellifer crops. For example, celery seeds of cultivar ‘Golden Spartan’ after initial imbibition at 18°C for 32 h were held at 1°C in aerated water for periods from 3 to 24 days,

Fig. 4.1. Bagged seeds following chitting treatment and/or in preparation for mixing with gel for fluid drilling. Note extensive radicle growth which is undesirable and likely to incur injury.

and then returned to 18°C and sown into peat blocks. Low temperature seed treatment of 24 days improved the uniformity of germination, and resulted in 79% germination as compared to 26% germination for the untreated seed (Finch-Savage, 1984). An earlier similar investigation with parsnip cultivar 'Harris Model' reduced the spread of time for germination, increased the percentage of seed germinating and reduced the time to germinate (Finch-Savage and Cox, 1982b).

Other seed treatments are also used to overcome seed dormancy. Robinson (1954) showed that at a constant temperature of 30°C or higher celery seeds require light for germination. Such dormancy is common in the *Umbelliferae* family, and it is known that the thermodormancy-release mechanism in celery is controlled by a red to far-red light reversible phytochrome system. The dormancy in celery seed imbibed in the dark can be broken by red light whereas the effect is reversed by far-red light. Seed of celeriac tend to have a stronger dormancy and require more light exposure for germination than seeds of celery (Pressman *et al.*, 1977). The response of celery seeds to light and chemical treatment is highly temperature dependent. At lower temperatures germination may proceed without light. Cold treatments at 1°C can bring about a partial change in seed hormone balance to overcome dark-induced dormancy in some species (Lewak and Rudnicki, 1977). Cytokinins and gibberellins have been implicated as mediating the effects of cold treatment (Thomas *et al.*, 1978a; Thomas and O'Toole, 1980).

The light requirement can be overcome by seed-soaking treatment at 5°C with a mixture of gibberellins GA_4 and GA_7 ($GA_{4/7}$) plus ethephon, or by osmotic priming in the light with polyethylene glycol (PEG) at 15°C (Brocklehurst *et al.*, 1982). Light seems to stimulate the production of gibberellins necessary for seed germination, and the higher the temperature, the greater is the need for light. However, light stimulation of germination can be negated by gibberellin (GA) biosynthesis inhibitors (Thomas, 1989).

Further mediation of the dormancy is improved when GA is combined with cytokinins or other compounds such as daminozide, and benzimidazole fungicides (Thomas *et al.*, 1975). Treatment with GA_4 and GA_7 increased the maximum temperature limit for germination but cytokinins raised this even further. The magnitude of this effect is cultivar-dependent and may be related to levels of natural inhibitors in the seeds. Additionally, seeds of cultivars responding to low concentrations of $GA_{4/7}$ appeared to contain less natural inhibitor than those requiring either high concentrations of $GA_{4/7}$ or cytokinin or a combination of both. Leaching experiments with celery suggest that other inhibitors, especially those that prevent germination of freshly harvested seed, probably occur in the outer regions of the seeds since they can be leached out very rapidly (Thomas *et al.*, 1975).

Seed priming is another conditioning treatment which permits water imbibition sufficient to initiate but not to complete germination. Priming is a procedure whereby seeds are placed in an aerated water solution containing specific concentrations of PEG or mineral salts such as K_3PO_4 or KNO_3 and other non-toxic materials that increase the osmotic concentration of the solution (Bradford, 1986). Thus the amount of water that the seeds imbibe can be controlled. Treatment conditions commonly provide solutions with an osmotic potential of −10 to −12 atm at 10–15°C for 1–3 weeks with continuous aeration. Partial water imbibition initiates the germination process, but further imbibition necessary to complete germination is restricted by the osmoticant.

Priming celery seeds in a PEG solution reduces mean germination time, increases percentage of seedling emergence and raises the upper temperature limit for germination (Brocklehurst and Dearman, 1983). Priming can also be used in combination with growth regulators to further enhance seed performance. Furthermore, it is claimed that some seed repair occurs during priming treatment. Following priming, the seeds can be dried back to the original moisture content and sown or stored for later use. Other priming practices use a solid matrix rather than a liquid and thus avoid the need for aeration. Many variables influence satisfactory post-treatment storage, which can range from several weeks to several months.

Assuming that temperatures are appropriate for growth, other measures to facilitate and advance emergence include the use of mulches or plastic film covers to enhance soil temperature and moisture conditions. Cultivation practices and materials such as phosphoric acid are used to avoid or alleviate soil crusting. Sowing at appropriate and uniform depths is important, although in some cases soil cover is not necessary. For some situations, uncovered seeds of celery can reduce the spread in emergence time and seedlings with larger cotyledons may result (Finch-Savage, 1986). Light bed rolling after planting is a practice that enhances seed-to-soil moisture contact. Using this procedure, Finch-Savage and Pill (1990) reported a higher percentage of carrot emergence, although this also resulted in lower seedling shoot weights. Following tillage and bed shaping, heavy drum rollers are used to firm the loosened peat soil in order to facilitate planting. Due to the resulting compaction, White (1978) found a linear relationship between depth of compaction and marketable yield; less compact beds had higher marketable yields. He showed that bed rolling following different tillage practices in Florida peat soils resulted in a significant increase of root forking compared to similar tillage without bed rolling. Thus, use of bed-rolling practices requires caution.

CARROT EARLY GROWTH

Root and hypocotyl growth

Soon after germination the taproot develops from the pro-meristem of the embryo and rapidly elongates, and with its fibrous lateral roots constitutes the root system. The storage roots of carrots and parsnips originate from the formation and activity of a cylindrical vascular cambium in the hypocotyl and taproot. At first, the vascular cambium consists of separate strips of cells formed from divisions of cells between the primary xylem and primary phloem. Soon thereafter secondary cambium development begins with meristematic activity between the primary xylem and phloem. It continues by extending around the xylem to form a complete cambial sheath around the central primary xylem. The cells of this cambium divide to produce daughter cells that form phloem tissue to the outside and xylem tissue to the inside. These cells expand and differentiate into vessels and mostly storage parenchyma. Root initials in the primary root give rise to the root cap and to the meristematic stele, cortex and epidermis which are initiated near the promeristem. The periclinal cambial divisions first become evident between the xylem and phloem. In the early enlargement of the storage root the cambium and the pericycle are fairly active in forming new tissues. In carrots, initiation of the secondary cambium usually precedes the development of foliage leaves (Esau, 1940).

With further development, the xylem and phloem parenchyma become meristematic with considerable carbohydrate accumulation and enlargement following. The enlarged storage organ is developed largely by secondary growth from the vascular cambium. In the fleshy storage organ of the carrot, parenchyma tissues which are permeated by vascular tissues predominate.

During secondary development of the storage root, the taproot apex continues to increase the length of the root, and lateral fibrous roots develop in both the secondary-thickened and primary regions of the taproot. Storage roots are characterized by cell division and expansion throughout their development. Lateral roots emerging from the endodermis are derived primarily from cells of the pericycle. The fibrous roots do not undergo secondary growth. Oil ducts in the intercellular spaces of the pericycle contain essential oils that are responsible for the characteristic aroma and various other flavour components of carrot (Senalik and Simon, 1987).

Continued enlargement of the storage root results in early shedding of the cortex tissues. The surface becomes covered and protected by the periderm that arises from the pericycle. Horizontal grooves that mark the exit of each lateral root appear on the periderm. When such markings are slight the root surface appears smooth. Wild carrots and primitive cultivars have pronounced lateral roots and root scars.

Storage root enlargement continues until harvest, albeit often at a slower rate later in the season. In the mature storage root, the central core of parenchyma is xylem-derived tissue separated from the outer sheath of phloem-derived parenchyma by a narrow layer of cambial cells and the whole is surrounded by peridermal tissue.

These taproot morphological and anatomical developments occur relatively early when compared to shoot canopy development, although shoot and storage root growth and development proceed together. Taproot elongation is very rapid and with favourable growing conditions often can achieve significant length in as few as 3 weeks after germination (White and Strandberg, 1978; Fig. 4.2). The depth of taproot penetration varies with cultivars, and may achieve a length of 1 m. The final taproot length is always longer than that of the enlarged storage root portion, and the distal portion of the taproot remains significantly thinner than the proximal or storage portion.

The terminus of the storage root portion is less clear in some cultivars such as 'Imperator' types, but obvious in others such as 'Nantes' types (Fig. 3.6). Stubbing is one of several terms used to identify the termination of storage root length. Good stubbing refers to a well-rounded storage root tip with only a thin remnant of the taproot remaining (Fig. 4.3).

The start of appreciable hypocotyl thickening and storage root enlargement usually commences after 6 weeks of growth. The hypocotyl region enlarges along with the upper portion of the storage root, and depending on cultivar type makes up the upper 15–20% of the final storage root length (Esau, 1940). The external hypocotyl/storage root distinction of the young seedling is gradually erased with the increase in root diameter and the development of fine fibrous lateral roots. The internal transition of root and stem structures also is gradual and indistinct.

Fibrous roots are absent on the upper hypocotyl portion of the storage organ, but an extensive number of fine, highly branched lateral fibrous roots usually grow from the mid and lower portions of the storage portion of the

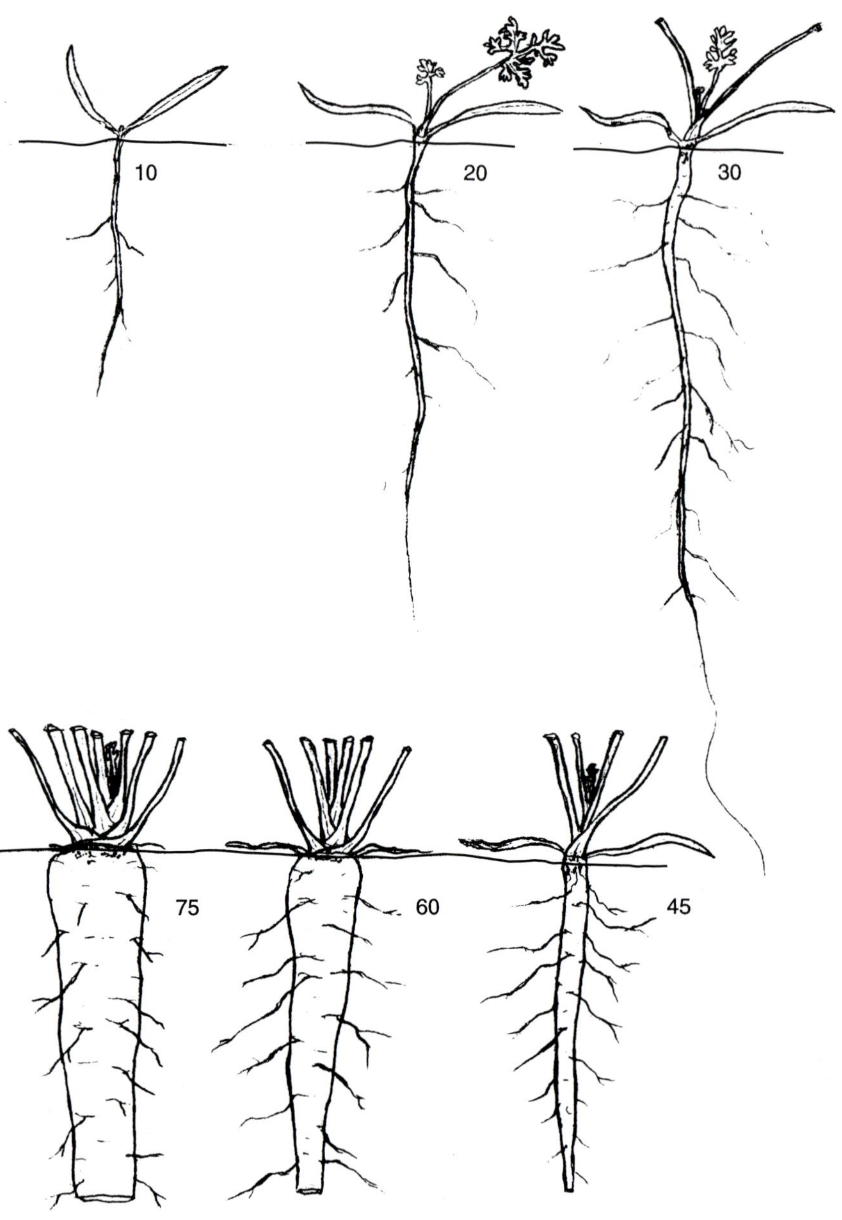

Fig. 4.2. Carrot storage root development at 10, 20, 30, 45, 60 and 75 days after planting.

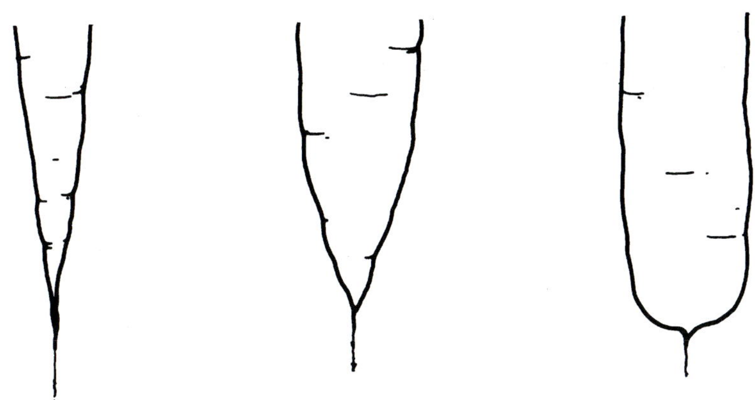

Fig. 4.3. Variation in carrot storage root tip shapes as influenced by cultivar, environmental conditions and/or developmental state. Far right root tip illustrates good stubbing.

taproot. These fibrous roots are usually concentrated within 30 cm of the soil surface.

An increase in early storage root length after emergence is favoured considerably by constant temperatures of 20 or 24°C versus 16 or 28°C, with 28°C being the least favourable (Fig. 4.4). Usually storage root length is rapidly achieved and is finalized by about 50 days after germination. That development is considerably more rapid than the gain in root weight. Root weight increase is very slow initially, but after the first third of the growth period root weight begins to increase steadily and enlargement continues until harvest. Near the harvest period the rate of gain in root weight declines. The rate of root diameter increase, as measured at the shoulder, occurs somewhat earlier than the rate of gain in root weight. Then the rate of increase in root weight parallels that of root diameter until the approach of harvest when the rate of diameter increase diminishes while the root continues to increase in weight. Storage roots that attain harvest stage or market-size status are often termed mature. In the true physiological sense they are not, since they are capable of further development. This is observed by the typical fill-in and increase in root diameter at the lower portion of the root compared to shoulder diameter.

Leaf length and weight achieve their full potential at about the mid-production period, whereas the potential for root diameter and weight is approached during the last quarter of crop growth. This pattern is representative of most cultivars, although occurring within a shorter time span for early developing cultivars. Total solids content (TDS), which reflects sugar accumulation, follows reasonably close to that of the root and haulm length pattern.

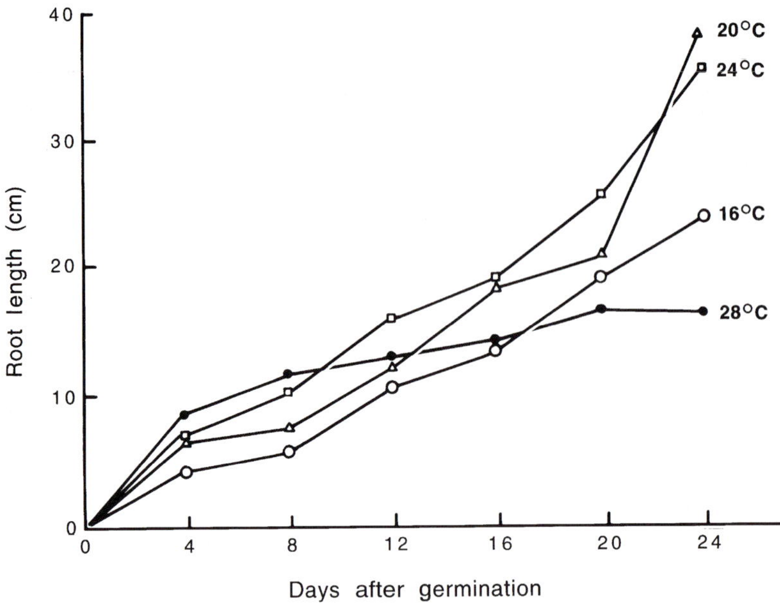

Fig. 4.4. Early linear carrot root development of plants grown in an organic soil under controlled environmental conditions with 12 h day length at four different temperatures. Source: White and Strandberg (1978).

Increasing carrot shoot and root temperature from 15 to 25°C increased shoot dry weight more than 2.5 times and storage root dry weight 1.5 times compared with maintaining shoot and root temperature at 15°C (Olymbios, 1973). Increasing shoot temperature from 15 to 25°C with root temperature at 15°C increased shoot dry weight (36%) and resulted in a slight (7%) reduction of storage root dry weight. The reversed situation that increased root temperature from 15 to 25°C while holding shoot temperatures at 15°C greatly reduced storage root dry weight (47%) while slightly (13%) increasing shoot dry weight.

Persistent low temperatures are associated with increased root length and more slender and cylindrical roots. Temperatures greater than 25°C, tend to restrict storage root length diameter and increase root taper (decreased cylindricality). High temperatures also can modify root shape by increasing sloping of root shoulders, and can slightly elevate the stem giving the appearance of a neck, whereas at favourable moderate temperatures root shoulders have a slightly rounded, nearly level or square-like appearance (Fig. 4.5). Preferred growing conditions for most carrot cultivars are warm day and cool night temperatures and this usually provides the best combination of root length and diameter. Overall,

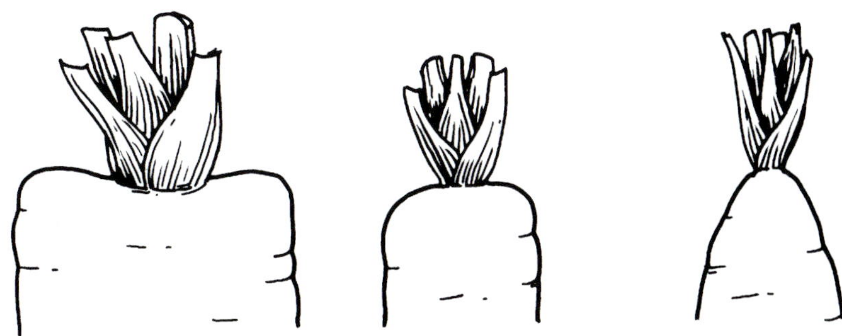

Fig. 4.5. Variation in carrot storage root shoulder shapes as influenced by cultivar, environmental conditions and/or development state.

genotype has the most influence on root shape.

Storage roots also tend to be longer in drier soils. The incidence of longitudinal root splitting is increased by large fluctuations in soil moisture. Root smoothness is also affected by non-uniform soil moisture conditions. When moisture supply is low the root surface may become dimpled and appear corrugated, whereas with prolonged excessive soil moisture, cork-like outgrowths may appear at lateral root bases or scars. Excessive soil moisture can also lead to increased lateral root formation so that the root surfaces appear hairy, and when harvested often have considerable amounts of soil adhering to the root.

Misshapen and forked roots may result from impediments to root development caused by debris from preceding crops, or by disease, nematode or other tissue injury. Soil compaction can restrict root length and alter root shape (Fig. 4.6). Plant density also influences root shape. High densities tend to reduce the taper of conical-shaped roots and increase the cylindricality of cylindrical-shaped roots. Low plant densities have an increased tendency to cause root splitting.

Longer growth periods also tend to increase the degree of cylindricality. Cultivars having a conical shape will exhibit a greater rate of enlargement near the shoulder early in growth, but as the growth period is continued, root diameter progressively broadens towards the tip, and rounding off at the tip (generally known as 'stubbing', 'blunting' or 'stumping') increases (Fig. 4.3). Stubbing or filling-in of the storage root tip occurs more rapidly in early production cultivars such as 'Nantes' type cultivars than those of late-developing cultivars like 'Chantenay', 'Danvers' and other conical-shaped and late-developing cultivars.

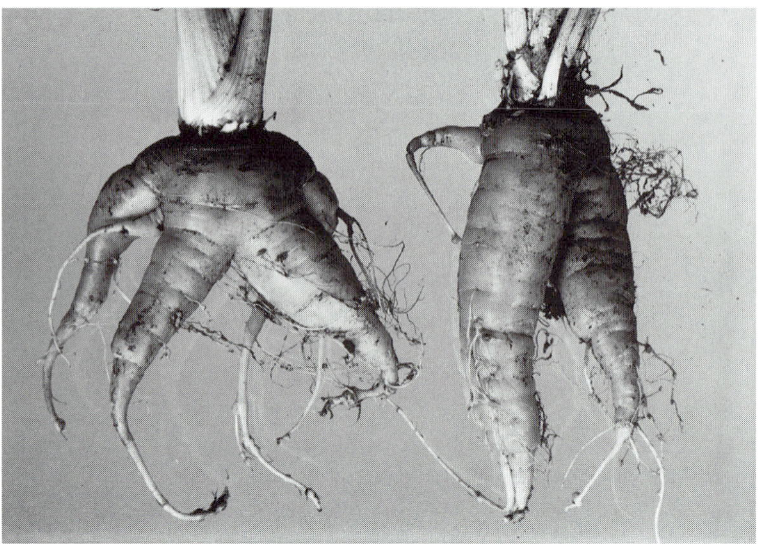

Fig. 4.6. Misshapen roots. The condition may be induced by *Pythium*, early nematode infection and/or abiotic causative factors.

Carrot leaf growth

Although initially relatively slow and short, leaf growth generally is upright and continuous, and new leaves are progressively longer until the time when rapid storage root enlargement begins. Concurrent with storage root enlargement the rate and length of new leaf formation declines. Bolting and flowering drastically slows leaf growth.

A reduced leaf growth rate and decline in new leaf formation may also be due to sub-optimal growing conditions and/or disease. Thus, the size and extent of foliage biomass is not always indicative of the extent of storage root development. It is not uncommon for situations to occur where small tops accompany large roots or large tops are coupled with small roots. This observation may be deceptive since the contribution made by earlier leaves, which are no longer present, to storage root enlargement is often overlooked. Even during similar growing conditions the ratio of foliage to root biomass can vary considerably between cultivars.

Temperature strongly influences early leaf length. Strandberg and White (1978) showed leaf length growth shortly after emergence and at constant temperatures of 20 or 24°C was almost twice that of plants grown at 28°C, and about three times longer when plants were growing at 16°C (Fig. 4.7).

A typical growth curve for a summer carrot crop is depicted in Fig. 4.8. Foliage length begins rapidly and at steady rate until about 60–70 days when maximum leaf length is achieved. Thereafter, the length of new leaf growth is

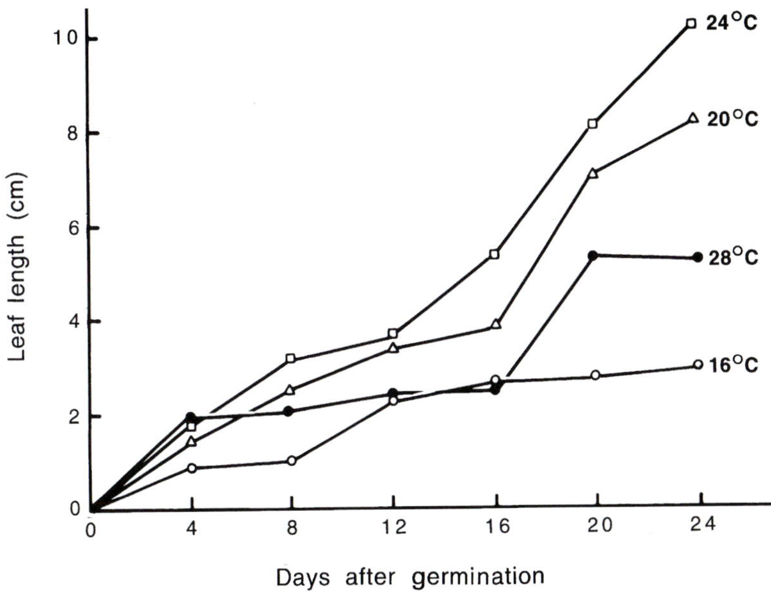

Fig. 4.7. Early linear carrot leaf length development of plants grown in an organic soil under controlled environmental conditions with 12 h day length at four different temperatures. Source: White and Strandberg (1978).

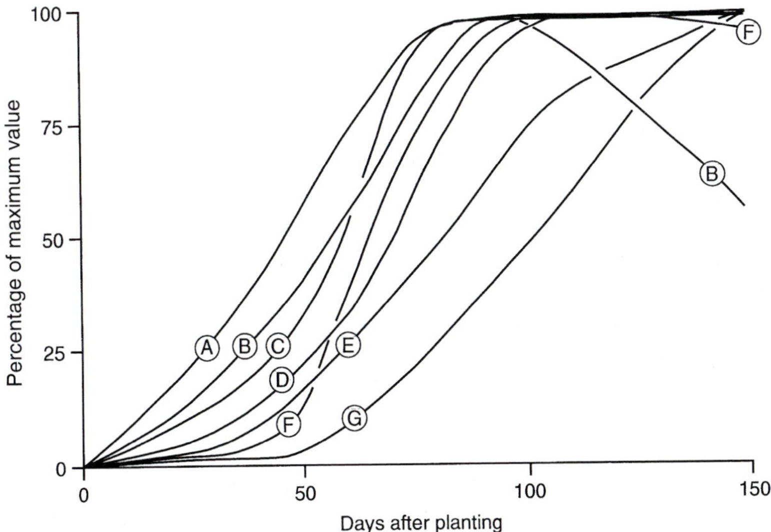

Fig. 4.8. Seasonal growth pattern of carrot leaves and storage roots grown as a summer crop in the mid-western United States (Wisconsin). A, total solids; B, leaf length; C, root length; D, carotene content; E, root diameter; F, leaf fresh weight; G, root fresh weight.

progressively less. However, the earliest formed leaves are smaller and weigh less until about 50 days after emergence when their size and weight rapidly advance. Subsequently developed leaves are heavier, but near the usual harvest period further leaf growth becomes smaller and weighs less. Overall, leaf length and weight achieve their full growth rate at or slightly after the plant's mid-growth period.

Leaf length, but not necessarily dry weight, tends to increase during long photoperiods and warm temperatures, and in high nitrogen, and high plant density conditions. Carbon dioxide enrichment increased the leaf area per plant, particularly during early root growth, primarily as a result of an increase in the rate of leaf area expansion, and not an increase in leaf number. At harvest, 137 days from sowing, total biomass was 16% greater at 551 p.p.m. CO_2 compared to 348 p.p.m. CO_2 (Wheeler *et al.*, 1994). Enrichment with CO_2 also increased root yield. Generally, larger leaf growth supports greater root enlargement, although high temperatures can increase respiration which can diminish storage root development. When nutrition, soil moisture and temperature are adequate, leaf colour will be darker than in situations where nitrogen and soil moisture are low and temperatures high.

CARROT STORAGE CARBOHYDRATES

The carrot vegetative shoot is the source of the products of photosynthesis, and also competes with the storage root as a sink for the photoassimilates. Transfer of the stored assimilates during bolting and flowering to the production of new tissue results in a decrease of dry weight and storage root sugar concentration. Thus, in a botanical sense, the primary function of the carrot storage root is that of a temporary reserve of photosynthates until utilized for development of the floral stem and inflorescence.

Photoassimilate distribution in carrots has recently been thoroughly reviewed by Hole (1996). Benjamin and Wren (1980) reported that fibrous roots regulate general growth and partitioning of metabolites most prominently in seedlings. Removal of fibrous roots of young carrots caused a considerable reduction in entire and relative shoot growth rate. Once secondary root development was initiated, about 10–21 days after germination, carbon assimilation movement was proportional to the size and activity of the sink (Warren-Wilson, 1972; Wareing and Partick, 1975). Storage root, fibrous roots and leaves compete more-or-less equally for a single assimilate pool with relatively little influence of factors like competition for light due to variability in plant density. Dry weight ratios, in contrast, are significantly affected by plant density (Hole *et al.*, 1983), although the partitioning of photoassimilates between storage root and leaves apparently does not involve much signalling between these sinks.

Carrot varieties vary significantly in storage root size and shape as well as in shoot size and architecture. Interestingly the shoot to root weight ratio early in development, about 27–48 days after germination, predicted partitioning of assimilates at plant maturity (Hole *et al.*, 1987). Under higher light intensity shoot weight was smaller, relative to that of storage root, than it was under lower intensities, but not due to a direct change in assimilate partitioning (Hole and Sutherland, 1990). Light intensity had no significant effect on relative sink activity in carrot (Hole and Dearman, 1993). Different photoperiod lengths (8 and 15 h) did not change shoot to storage root weight ratio, although total plant dry matter was increased for plants receiving a longer day length (Olymbios, 1973).

Based upon evaluations of respiratory and photosynthetic rates and in carbon partitioning, sink activity of fibrous roots in early development (20 days after planting) of larger-rooted carrot cultivars exceeded that of small-rooted cultivars (Hole and Dearman, 1991) while respiratory and photosynthetic rates were similar. Photosynthetic gain and respiratory loss during storage root initiation (14–28 days) were examined relative to development for cultivars 'Kingston' and 'Super Sprite' that have a high and low shoot to storage root ratio at maturity, respectively (Steingrover, 1981; Hole and Dearman, 1991). Root respiratory losses in each study were similar for each cultivar, about 17% of daily photosynthesis. The proportion of assimilate allocated to the shoot (64%) and total root (36%) were also similar for the two cultivars. Interestingly, the proportion allocated very early to the young storage root was 1.6 times greater in cv. 'Super Sprite'. Hole (1996) suggests that it may be the initial distribution with the root system rather than between shoot and total root that determines cultivar differences. Pruning of the taproot and fibrous roots indicated that either of these alterations of sink size had some influence on assimilate partitioning but the relationship was not very strong (Benjamin and Wren, 1980).

The transport of photoassimilate from leaves to roots, compartmental analysis in sinks of carrot and overall assimilate flux have not been studied extensively in carrot, and clarification of these processes is made difficult by tissue complexity, intermingling of source–sink and primary–secondary tissues, and complications of floral initiation in late growth.

Exogenous application of a growth regulator like gibberellic acid (GA) increases the ratio of dry matter distribution between carrot foliage and storage root, suggesting that endogenous hormones regulate dry matter partitioning. Application of chlormequat chloride, an anti-GA compound, decreases the ratio (McKee and Morris, 1986).

Glucose and fructose accumulation predominates early in storage root development, when invertase activity is highest, while sucrose accumulation predominates later in development. However, the ability of carrot storage root parenchyma to store sucrose develops within a few days of its formation and the concentration increases with time. About two-thirds of the mature storage root total sugar is sucrose, but depending on stage of plant development, cultivar and environmental difference the level will vary (Phan and Hsu, 1973). Carrot cultivars differ considerably in their ability to accumulate sugars in the storage root. Those with the capacity for high sugar yield have higher net assimilation rates (Lester *et al.*, 1982). Generally, phloem parenchyma has a higher concentration of total sugar per unit fresh weight than xylem parenchyma, and the ratio of sucrose to hexose is also greater in phloem than xylem tissue (Phan and Hsu, 1973).

Future studies examining the interplay between gene activation, enzyme activity, environment, nutrients and growth regulators should broaden our understanding of carrot photoassimilate partitioning and carbohydrate storage for utilization in quality and yield improvement.

CELERY LEAF GROWTH

Consisting of a short crown stem and rosette leaves, the yield of stalk celery is maximized by the production of many leaves with long and thick petioles. The rate of leaf development is temperature-dependent, and both low- and high-temperature situations retard leaf formation. Long days or night breaks also cause reduced leaf production (Roelofse *et al.*, 1990).

Long days extended artificially or by night breaks caused a reduction of leaf number, but leaves were elongated. Under short days leaves were shorter (Roelofse *et al.*, 1990). Leaf elongation is also affected by high temperature; celery grown during high temperature has shorter petioles. The influence of high temperature and short days can be reversed with gibberellic acid treatment (Pressman and Negbi, 1987).

CELERY AND OTHER UMBELLIFER STORAGE CARBOHYDRATES

Unlike other C3 plants, the main products of CO_2 fixation by photosynthesis in celery are sucrose and the sugar alcohol mannitol. It has been estimated that mannitol forms as much as 50% of the photoassimilate translocated in the phloem. The remainder of the photoassimilate is mostly sucrose. Sucrose is produced and utilized in leaves of all ages, whereas mannitol is primarily synthesized in mature leaves, utilized in young and stored in all leaves. Both of these carbohydrates are produced in the mesophyll and transported in the phloem of the leaves. The petioles mainly store mannitol, glucose and fructose, but very little sucrose.

Mannitol derives from mannose by action of the NADPH-dependent enzyme, mannose-6-phosphate reductase. Mannitol is stored mostly in the vacuoles of the parenchyma cells in the petioles and is broken down to CO_2 only in actively growing tissues. Photosynthetic rate in celery is exceptionally high for a C3 plant, being in the order of approximately 60 g CO_2 dm^{-2} per hour. This may be due in part to the use of excess reductant produced photochemically in mannitol synthesis, but most likely the high rate of photosynthesis is a dividend gained in plants forming mannitol as a photoassimilate. The advantage of mannitol over sucrose is mostly energetic, since being an alcohol, it is more easily reduced than sucrose. Therefore, the complete oxidation of mannitol will produce more energy per mole than sucrose. Polyols, such as mannitol, have been found to accumulate in plants in response to environmental stress including salinity. Since celery is tolerant to salt, mannitol may be responsible for this attribute, serving as an osmoprotectant. Growing celery plants in the presence of NaCl results in high accumulation of mannitol in the tissues, probably to alleviate stress in plants (Pharr *et al.*, 1995).

Another unusual sugar which occurs in vegetable umbellifers is apiose. Found in the cell wall of parsley, apiose is found in warm water extracts. Its role in parsley quality has not been examined.

Parsnip accumulates starch as a storage carbohydrate. Starch metabolism in parsnip has not been studied. Sweetness is generally found to increase for parsnip very late in the season, especially after frost to which is attributed a conversion of starch to sugars.

CAROTENOIDS

Carotenoids are widespread and important natural pigments found in higher plants, algae, fungi and bacteria. Carotenoids are found in chromoplasts and also in chloroplasts and non-green plastids. In carrot roots, chromoplasts develop directly from proplastids. The chromoplasts are the crystal-containing type; the carotene crystals form when a critical concentration is reached. The major carotenoids found in carrot roots and foliage include: β-, α-, γ-, and ζ-carotene; several others have a minor presence (Simon and Wolff, 1987; Fig. 9.1).

One of the most important physiological functions of carotenoids is to act as vitamin A precursors in animal organisms. About 40 carotenoids are known to be vitamin A precursors. β-Carotene, present in almost all vegetables, is the most widespread pro-vitamin A carotenoid. Alpha- and β-carotene are the major pigments responsible for orange and yellow carrot root colour. β-Carotene often may represent 50% or more of the total carotenoid content, and usually constitutes twice the content of α-carotene. Lycopene is the predominant carotenoid pigment responsible for red colour, and anthocyanin is the primary pigment in purple-coloured roots. The anthocyanin pigment is derived from a different metabolic pathway, and is not a carotenoid. Xanthophylls are oxygenated derivatives of carotenoids and are accessory pigments present in small amounts in orange carrots; they are much more prominent in yellow carrots. White carrots do not develop pigmentation.

Carotenoids are not uniformly developed or distributed in the root. Carotene synthesis is more advanced in the more mature tissues so that these tissues typically contain more carotene than more recently formed tissues, and thus the concentration of pigment decreases longitudinally from the upper portion of the root to the root tip. It is also typical that phloem tissues have a higher concentration than xylem tissues.

With the increase in plant age and size, carotene accumulates and root colour changes. As growth and carotene synthesis continue, root colour intensifies and the colour gradient between tissues and along the root length also appears to be less pronounced.

Carotene synthesis initially is very slow until about the middle of the growth period and becomes more rapid during the middle to latter period of the plant's growth. The rate of carotene development and accumulation is more rapid in early-developing cultivars than for those that require a longer growing period for development. Carotene development and accumulation continues during growth, and for most production situations is at or near maximum about the time of harvest. About this time carotene synthesis generally tends to level off or may begin a slight decrease (Pepkowitz *et al.*, 1944). However, if favourable temperature and moisture permit growth to continue, carotene synthesis may also continue, although at a reduced rate. Carrot producers occasionally delay harvests in order to obtain increased root colour.

The content and rate of carotene change vary with carrot cultivars and environmental factors (Buishhand and Gabelman, 1979; Simon *et al.*, 1982). Simon and Wolff (1987) also showed cultivars varied in individual and total carotenoid content with regard to genotype, location and season. Although differences in carotene levels occurred when the same cultivars were grown in different locations and soil types, genotype overall was the primary determinant of maximum achievable storage root carotene content (Table 4.1). Extensive studies of root colour inheritance in carrots have been carried out in order to improve cultivar carotene content (Buishhand and Gabelman, 1979). The United States Department of Agriculture (USDA) carrot breeding programme in 1970 initiated the selection for dark orange roots which resulted in substantial increases of carotene content in many cultivars and breeding stocks (Simon, 1987). Selective breeding has increased total carotene content in US grown cultivars from 50 to 70 p.p.m. before 1970, to between 70 and 120 p.p.m. in the late 1970s and early 1980s, with many cultivars presently at about 150 p.p.m. Breeding populations with 200–300 p.p.m. carotenoids are relatively common and several populations with 400–600 p.p.m. have been developed (Simon *et al.*, 1990). As previously mentioned, growing conditions do influence carotene content, and thus relatively mature roots are needed to assure expression of maximum genetic potential.

Table 4.1. Carotene concentration of several carrot cultivars and germplasm.

Cultivar/germplasm	Total carotene content (mg g^{-1} fresh weight)
'Chantenay'	41
'Nantes'	59
'Hi Colour 9'	65
'Danvers 126'	71
'Imperator 58'	78
'A Plus'	168
'Beta III'	270
High carotene mass (HCM) selection	475

Source: Simon (1987).

Environmental conditions having a role in carotene development and accumulation include high soil moisture which tends to reduce root carotene content. Unless deficient, fertility generally has little effect, although Evers (1989b), working with carrots in Finland, reported that different fertilization application practices could produce a higher carotene content, although the gain obtained was relatively small. Carrots grown on muck soils that were slightly more acidic had high levels of carotene (Simon *et al.*, 1982). Carrots grown in sandy soils often have slightly higher carotene levels. Both high and low temperatures, greater than 30°C and less than 5°C, respectively, greatly reduce carotene synthesis, which has an optimum between 15 and 21°C. A regime of cool night and warm day temperatures favours good colour development.

In some situations the initial carotene content of storage roots may decrease at the beginning of storage. In some good storage conditions carotene content is either maintained and/or slightly increased up to the time sprouting occurs. Carotene biosynthesis apparently occurs even during storage as carotene levels were observed to increase during a storage period of 20 weeks at temperatures between 0 and 4°C (Lee, 1986). Root moisture loss may account for some of the observed increase.

BOLTING AND FLOWERING

Carrot

At the beginning of carrot floral development the growing point of the seed stem changes from slightly convex and becomes more cone-shaped as the main axis of the seed stem elongates (Borthwick, 1931). As growth and elongation continues, the apex of the flowering apex begins to flatten. Around its periphery the primordia of involucel bracts appear as small protuberances. As these increase in size they are followed centripetally by umbellet primordia, and the entire stem apex is covered with these primordia, but in different stages of development. Sepal, petal and stamen primordia occur at almost the same time. The stamen primordia are above those of the sepals. Petal primordia soon exceed in size the sepal and stamen primordia. Carpel primordia are the last to develop. In each carpel the cavity in the ovary divides into two locules. Early in plant growth, even before substantial top foliage growth occurs, the primordia of the umbel and umbellets are well differentiated.

Flowers open irregularly. The filaments are in-rolled in the unopened flower, and at anthesis the petals separate slightly, the anthers are released and the filaments straighten. The pollen is shed and most of the stamens fall before the stigmas are receptive. Style elongation and flower opening occur at about the same time, but fertilization occurs after the styles have separated and just before petals fall.

The effect of low temperature that induces floral stem initiation is known as vernalization. The stage of growth when carrots seedlings are not responsive to low temperature vernalization is known as juvenility. That condition usually ends when carrot plants have initiated 8–12 leaves, and storage roots are greater than 4–8 mm in diameter. The level of response is cultivar-dependent. Atherton *et al.* (1990) reported that the thermal time required from sowing to complete juvenility for cv. ‘Chantenay Red Cored’ was 756°C days. Meeting the subsequent minimum requirement for vernalization was 126°C days. Temperate-region carrot cultivars are usually vernalized by exposure for 2–8 weeks at constant temperatures between 0 and 10°C (Dickson and Peterson, 1958).

Chilling treatments of 11–12 weeks at 5°C applied to carrot plants (cv. ‘Chantenay Red Cored’) maintained in darkness or photoperiods of less than 12 h resulted in more rapid and prolific flowering than chilling under longer photoperiods (Craigon *et al.*, 1990). The rate of flower bud appearance and rate of stem internode extension were linearly related to temperature. They also reported that the elapsed time from the end of vernalization to the start of visible bolting and flowering decreased with increasing thermal time of vernalization. As temperatures increased from –1°C to an optimum of about 6°C the rate of elongation increased, but the rate decreased as temperatures were increased to a maximum of about 16°C.

Leaf area and fresh and dry weights were greater after chilling in plants that had received the long photoperiod. Chilled plants returned to long photoperiods (16 h) in a warm glasshouse (not less than 16°C) flowered, whereas those returned to short photoperiods (8 h) in otherwise similar conditions remained vegetative (Atherton *et al.*, 1984). It was also shown that photoperiod before or during chilling has a considerable effect on the flowering response to chilling. Short photoperiods (12 h) or darkness during chilling resulted in an earlier and a more promotive effect on flowering than long photoperiods (16 h). Long days after vernalization stimulated flowering (Atherton *et al.*, 1984). After vernalization, flowering could be suppressed with continuous low light.

These results with young carrot plants conflict with earlier indications of Dickson and Peterson (1958) that carrots are day-neutral after chilling. Atherton *et al.* (1990) concluded that carrots are 'short-long day plants' with a chilling requirement for flowering similar to celery.

In addition to stimulating flowering, short photoperiods or darkness reduced the growth of leaves during chilling, which may be related to flower induction either as a cause or effect. The rate of bolting and flowering for cv. 'Chantenay Red Cored' increased linearly with temperature from –1 to 5°C, and declined linearly with temperatures from 7 to 16°C. This relationship permitted determination of base, optimum and maximum temperatures for vernalization of this cultivar as –1, 6.5 and 16°C, respectively (Atherton *et al.*, 1990).

The results of experiments by Atherton *et al.* (1984) and Craigon *et al.* (1990) enabled them to develop a model that closely predicted the bolting and flowering response of carrot (cv. 'Chantenay Red Cored') following accumulation of various thermal-times of vernalization in controlled environments and different field conditions.

Physiological changes occur during vernalization before any morphological changes are evident. Endogenous gibberellic acid (GA) levels rise as a response to cold treatment which stimulates the flowering process. Exogenous application of gibberellins has successfully induced flowering in carrot, although this technique is not widely used.

Plants which have been induced to flower, but are not yet undergoing stem elongation can be de-vernalized by a few days of high temperature (28–35°C). This occasionally occurs when vernalized roots are planted in a warm glasshouse. De-vernalization can be reversed with further cold treatment. Once plants are reproductive their growth rate increases at temperatures from 14 to 26°C. Such plants mature sooner at the higher temperature with the result that the final size of plants is lower at the higher temperature. Concomitant with changes of the above-ground plant portions during bolting, carrot root physiology also changes dramatically. Although not well characterized, clear changes in carbohydrate metabolism occur during bolting since the xylem quickly becomes very lignified before the floral stalk elongates. This textural change renders the root tissues inedible.

Celery

Celery bolting and flowering are dependent on the relationship of genotype and environmental conditions that include temperature, photoperiod and light quality. Bolting and flowering are independent but closely associated developmental stages which often appear to occur concurrently. Bolting is a preliminary stage in the transition from vegetative to reproductive growth in biennial rosette-type plants. Stem elongation occurs only in vernalized celery plants whose apices have differentiated into flower primordia (Pressman and Negbi, 1980). Low temperature-induced bolting can be replaced by exogenous GA treatments, but flowering seldom occurs after such treatment (Pressman and Negbi, 1987).

Zeevaart (1978) suggested that the developing floral meristem was a source of endogenous gibberellins that control stem elongation rather than having a direct role in the induction, and that the rise in the level of GA-like substances following vernalization is a result rather than the cause of flower formation. Plants in the vegetative growth phase have an almost flat (Fig. 2.11a), relatively large diameter apex of about 0.4–0.5 mm. As plants enter the reproductive phase, the apex becomes pointed as it rises (Fig. 2.11b) and decreases in diameter. The major indication of bolting is the elongation of the inflorescence stalk which can be monitored by measuring changes in the diameter of the stem apex.

Celery plants generally will not bolt in the first year of growth if non-inductive temperatures persist. Chronological age is not a reliable indicator of plant responsiveness to vernalization induction. Rather it is the physiological age along with the appropriate temperature that determines the response to vernalization; that physiological age varies among cultivars. Some cultivars are more easily induced to bolt than others, a distinction well known to celery growers. However, the duration of juvenility for some plants may extend until they have 17 leaves or more (Table 4.2). The observation of bolting with the occurrence of high temperatures occasionally leads to the erroneous implication that high temperatures cause bolting, when in fact high temperatures simply accelerate and amplify the expression of the already induced bolting.

In general, vernalizing temperatures of 5–10°C are satisfactory, although for some cultivars temperatures up to 14°C can be adequate for thermo-induction (Roelofse *et al.*, 1990). The duration of chilling temperature is cumulative and full induction usually requires several weeks. The lower the temperature and the longer the cold exposure period, the earlier and more rapid flower initiation and bolting will appear after removal from the cold treatment. Low temperature effects on flowering and bolting are more pronounced in older plants.

Sachs and Ryski (1980) reported pre-devernalization of nursery-grown celery transplants by exposure to high temperatures (25–35°C) for short periods before transplanting into the field caused a significant delay in the rate of bolting. Lengthening the period of high temperature exposure further delayed the onset of bolting.

Chilling imbibed seeds of celery cv. 'New Dwarf White' for 6 and 8 weeks

Table 4.2. Effects of 9 weeks of chilling at 5°C on bolting and flower initiation in celery cultivars 'New Dwarf White' (NDW) and 'Celebrity' (CEL). Bolting was recorded as having occurred when stem length exceeded 40 mm as measured from stem base to stem apex.

Plant age at start of chilling (weeks after sowing)	Leaf no. at start of chilling		Plants bolting and flowering at harvest (%)	
	NDW	CEL	NDW	CEL
3	4	2	0	0
5	8	7	0	0
7	17	17	100	91
9	19	21	100	100
11	24	25	100	100
13	28	29	100	100

Source: extracted from data of Ramin and Atherton (1991b).

at 5°C induced flowering in about half of the plants compared with non-treated seed when each were subsequently grown at a minimum temperature of 15°C. An 8-week cold treatment induced flowering at a lower leaf number than a 6-week treatment. Although chilling during germination had a positive vernalization effect for cv. 'New Dwarf White', the effectiveness depended on post-chilling treatment (Ramin and Atherton, 1991a). Plants from chilled seed subsequently grown at 20°C remained totally vegetative, indicating that growth at 20°C has a de-vernalization effect. Chilled imbibed seed of another cultivar, 'Celebrity' exhibited no vernalization response.

Chilling of mother seed plants at different stages of seed development had no influence upon flowering or bolting of progeny of seed from chilled seed plants compared to untreated seed-producing plants. Examination of shoot apices of treated and untreated plants showed all were vegetative (Ramin and Atherton, 1991a).

Photoperiod and light level conditions prevailing before, during and after vernalization chilling can influence flowering and subsequent shoot extension.

Before chilling

Holding plants in darkness for 4–8 days at 20°C just before chilling resulted in a significant delay to bolting and reduced the number of plants flowering. Generally, as the duration of darkness increase of for plants confined to darkness, subsequent leaf number, leaf area and shoot weight decreased. A shorter period of darkness had no such effect.

During chilling

Pressman and Negbi (1980) found long photoperiods during vernalization inhibited flower initiation and bolting in celery, whereas after vernalization long photoperiods promoted both. Thus they reported celery to be a quantitative short–long day plant with a vernalization requirement. Short photoperiods during chilling generally advance flower initiation and bolting. Total darkness during chilling even after 9 weeks at 5°C completely prevented any subsequent vernalization response either as bolting or as flowering, which is in contrast to vernalization of carrots where darkness during chilling enhanced subsequent bolting and flowering (Ramin and Atherton, 1994). Reduced light level during chilling had no effect on vernalization response except that a very low level of light slightly increased the time to bolting, and greatly reduced the percentage of both bolting and flowering plants. The importance of light during vernalization may depend on the level of stored carbohydrates in the shoot. Plants with high levels of shoot carbohydrate may be more sensitive to light during chilling, but those with low reserves may not (Ramin and Atherton, 1994).

Night breaks

Regimes of long photoperiods with light interruption (night breaks) during vernalization reduced bolting and flowering in celery and decreased stem length. Strong retarding effects of nightbreaks on flowering and bolting were observed at 10°C. Since all plants induced at this temperature flowered, it is clear that the photoperiod for short days during vernalization is only quantitative (Roelofse *et al.*, 1990). Since flowering was delayed when either long days or long nights interrupted by night breaks are given during vernalization, night break lighting can be a useful procedure for limiting bolting in early celery crops grown at cool temperatures under glass (Roelofse *et al.*, 1990).

After chilling

After 9 weeks of chilling at 5°C, long photoperiods promoted and short photoperiods decreased the proportion of both bolting and flowering plants. After chilling, no significant enhancement of bolting and flowering were observed with increased irradiance (Ramin and Atherton, 1994).

Flowering, meiosis and pollination ensue in the 1–3 months of plant growth after vernalization. The rate of floral structures and seed development is primarily determined by temperature, but genetic factors also play a significant role. For breeding programmes it is desirable that breeding stocks produce seed in as short a time as possible after planting vernalized stocks. Therefore, breeders have selected stocks which do not flower prematurely, but which do flower and set seed in sufficient time to keep the seed production crop within an annual cycle.

5

CROP PRODUCTION

ENVIRONMENTAL INFLUENCES

Climate

Climate has the greatest overall influence on crop production, and temperature is its most important component. Most umbellifer vegetables fit the category of cool season crops and thus are generally produced in temperate climates. When produced in low latitude regions, they typically are grown at high elevations for the advantage of moderate to cool temperatures. The optimal mean growing temperatures are fairly similar for most umbellifers and commonly range between 15 and 21°C (Table 5.1). Mean temperatures less than 10°C or greater than 25°C tend to limit growth and eating quality, particularly those umbellifers grown specifically for their foliage. Quality characteristics of most edible storage root crops also benefit from slightly cooler conditions (10–15°C). Diurnal temperature fluctuations providing moderate day and relatively low night temperatures tend to improve carbohydrate accumulation in storage roots. At temperatures much above 25°C the increased rate of plant respiration tends to limit root yields.

Crops with a greater adaptation to warmer temperatures in the 21–25°C range include fennel and coriander. Moderate temperatures (15–21°C) most benefit carrot, celery, parsley, parsnip and chervil. The lower portion of this temperature range is preferable for lovage, Japanese hornwort and water dropwort growth. Parsnips, parsley and carrots are fairly tolerant of low temperatures and have some frost tolerance. Parsley is a hardy crop that overwinters well, and can tolerate short periods of fairly low or high temperatures. A period of low temperatures and even light freezes prior to harvest actually improve parsnip root eating quality because of root starch conversion to sugars. Most other umbellifer crops, notably celery and arracacha, are frost-sensitive. Extended periods below 10°C can vernalize and

Table 5.1. Optimum soil and environmental conditions for several vegetable umbellifer crops.

Crop	Temperature (°C) range / optimum	Soil			Light	Nutrition requirement
		Type	Moisture	pH		
Angelica	5–20 / 15	Loam, clay loam	Moist	6.0	Full sun, light shade	Moderate to high
Anise	8–25 / 20	Loam, sandy loam	Well drained	6.5	Full sun	Moderate
Arracacha	10–25 / 18	Loam, sandy loam	Well drained	5.5	Full sun	High
Caraway	6–25 / 18	Sandy loam, loam	Well drained	6.4	Full sun	Moderate
Carrot	5–35 / 18	Loam, sandy loam	Moist, well drained	6.0	Full sun	Moderate to high
Celery	10–30 / 18	Loam, clay loam	Moist, well drained	6.0	Full sun	High
Celeriac	10–30 / 18	Loam, clay loam	Moist, well drained	6.0	Full sun	High
Chervil	7–21 / 15	Loam	Moist well drained	6.5	Partial shade	Moderate
Coriander	7–27 / 20	Loam, sandy loam	Well drained	6.6	Full sun high	Moderate to
Cumin	9–26 / 17	Sandy loam	Well drained	6.0	Full sun	Moderate
Dill	6–26 / 20	Loam	Moist, well drained	6.0	Full sun	Moderate to high
Fennel	4–27 / 20	Loam, wide range	Moist, well drained	6.5	Full sun, light shade	Moderate to high

Lovage	6–18 / 15	Loam	Moist, well drained	6.5	Full sun, light shade	Moderate to high
Parsley	5–26 / 20	Loam	Moist, well drained	6.0	Full sun	Moderate to high
Parsnip	10–35 / 18	Loam, sandy loam	Moist, well drained	6.0	Full sun	Moderate to high
Smallage	10–30 / 18	Loam, clay loam	Moist, well drained	6.0	Full sun	High
Japanese hornwort	10–20 / 15	Loam, clay loam	Moist	6.5	Light sun shade	High

induce premature bolting of celery, celeriac, carrot, parsnip and parsley plants. On the other hand, annual umbellifers may prematurely bolt when exposed to periods of high temperature.

Because carrots have a broad temperature tolerance, production is feasible throughout the year in some locations. Mean day temperatures between 15 and 21°C are considered favourable for carrot root and top growth, and at these temperatures root colour and shape are also optimized.

Continuous mean temperatures above 25°C or less than 15°C tend to limit carrot colour development. Roots grown during cool nights (7°C) alternating with moderate days (15°C) develop more carotene than roots maintained at a continuous 7°C. Simon and Wolff (1987) reported significant differences with regard to carotene content occur between carrot cultivars in their sensitivity to growing temperatures and locations. At temperatures greater than 25°C, top growth is enhanced preferentially to root growth. Temperatures greater than 30°C tend to limit top growth, and when prolonged, also result in undesirable root colour, flavour and textural changes. However, 'tropical' carrot cultivars tolerate and can be productive at mean temperatures between 25 and 30°C, but also tend to be more sensitive to low temperatures and more susceptible to early bolting.

Temperatures also influence carrot root length and diameter. At mean temperatures of 12–13°C, roots tend to grow relatively long and slender, whereas at a constant 24°C, roots are shorter and thicker. Alternating low night (7°C) and moderate day (18°C) temperatures also tend to produce roots that are long and slender compared with those grown at a constant 18°C or higher. Carrot foliage and roots usually can tolerate periods of low temperature or light frost with little or no apparent injury.

Celery and celeriac are often grown as an autumn or spring crop where summer temperatures are high, and as a winter crop in regions with warm or moderate temperatures. When directly sown, crops are harvested after about 5–6 months of growth. When grown from transplants, crops are harvested after 3–4 months. Moderately warm and long sunny days are ideal for high quality and yield. Temperatures above 25°C adversely affect quality while temperatures less than 10°C suppress growth and may stimulate bolting.

Umbellifer crops differ in their sensitivity to light intensity, quality and duration. Some are more productive in full light, whereas others perform better with partial shade. The light environment can also influence crop quality, particularly with regard to essential oil composition and the content of flavour and aromatic compounds produced in leaf and root tissues and seeds. Seeds of celery, celeriac, chervil, parsley, parsnip and carrot usually germinate at a higher rate and more rapidly in the light than in the dark, especially during higher temperatures. However, daily alternating temperatures during celery germination can replace the light requirement (Pressman *et al.*, 1977). Gibberellin and cytokinin treatments also effectively replace the need for light in celery germination (Thomas *et al.*, 1975). Tanne and Cantliffe

(1989) improved germination percentage and growth rate with celery cv. 'Earlybelle' 20 days after treatment at 15 or 25°C when seed were pretreated with polyethylene glycol (PEG) priming for 14 days at −12.5 bar. Germination was also enhanced by the addition of growth regulators, especially cytokinin to the priming solution (Table 5.2a). They also reported that differences in cultivar response were observed (Table 5.2b).

Prevailing climate is outside of grower control. Even so, producers are ingenious in effecting some degree of climate modification with their use of field orientation, irrigation, mulching, shading and wind barriers.

Soils

Field location and soil conditions play a major role in the successful production of umbellifer vegetables and vegetable producers tend to utilize the most productive soils available in order to obtain high quality and yield. Level fields with good drainage are preferred because they require less management, and are less subject to erosion. Sloping fields can be satisfactorily cultivated, although contour or terrace cultivation may be required. In some situations it may be necessary to level field surfaces in order to flood or furrow irrigate. Another example is levelling peat soils to accommodate the periodic flooding of fields for disease and pest eradication, and retard peat loss from oxidation.

Most vegetable umbellifers can grow satisfactorily in a wide range of soils. Ideally, soils should be deep, friable, fertile and relatively high in organic matter. Uniform soils, in good physical condition with a soil particle structure providing good nutrient and water holding capacity, and free of compacted layers, are desired attributes. Good drainage characteristics and an absence of salinity are other obvious features.

Sandy and light sandy loams are desirable where early cropping is important. Such soils usually are better drained and aerated, warm rapidly, and can be tilled relatively soon after rain or irrigation with less compaction and soil texture damage. However, light-textured soils usually have low moisture retention, often a low plant nutrient content, and are more easily leached. Accordingly, they require additional fertilizer and moisture to achieve high yields. Light-textured soils are preferred for fresh market carrots in order to facilitate harvesting, and to produce smooth root surfaces.

Although less favourable for early production of vegetable umbellifers, loam and silt loam soils typically are more fertile and have a greater potential for high yields. These soils generally have good moisture-holding properties, are less subject to leaching, and thus, tend to better retain plant nutrients.

Fine-textured clay soils are usually less adapted to intensive vegetable umbellifer production because of possible aeration and drainage limitations that may restrict nutrient acquisition and root development. Their tendency to retain moisture can make tillage and seedbed preparation more difficult. Being slower

Table 5.2a. Percentage germination after 20 days and mean days to germination (MDG) of celery seeds cv. 'Earlybelle' at 15 or 25°C as influenced by priming treatments at 15°C for 14 days with growth regulators.

	Germination temperature			
	15°C		25°C	
Treatment	% Germination	MDG	% Germination	MDG
Untreated seeds	66a*	11.4d	20c	15.4d
PEG at −12.5 bar	61ab	6.7ab	27b	9.1b
PEG + BA (100 p.p.m.)	50b	11.8d	43a	14.6d
PEG + $GA_{4/7}$ (100 p.p.m.)	63ab	8.2c	11c	14.6d
PEG + ethephon (500 p.p.m.)	59ab	6.6ab	20bc	9.5b

*Values in each column are significantly different from each other by Duncan's multiple range test at 5%.
PEG, polyethylene glycol; BA, benzyladenine; GA, gibberellins.
Source: Tanne and Cantliffe (1989).

Table 5.2b. Percentage germination in the dark at 25°C, radicle length and hypocotyl length of celery seedlings 24 days after treatment as influenced by priming treatments in −12.5 bar polyethylene glycol (PEG) at 15°C for 14 days with growth regulator combinations.

	Cultivar					
	'Earlybelle'			'683K'		
Treatment	Germination (%)	Radicle length (mm)	Hypocotyl length (mm)	Germination (%)	Radicle length (mm)	Hypocotyl length (mm)
Untreated seeds	0b	—	—	0b	—	—
PEG at −12.5 bar	0b	—	—	26b	9.5a	27.0ab
PEG + $GA_{4/7}$ + BA (100 p.p.m. + 100 p.p.m.)	22a*	4.5a	11.5b	82a	4.0b	10.5c
PEG + $GA_{4/7}$ + ethephon (100 p.p.m. + 500 p.p.m.)	2b	6.0a	11.0b	2b	5.5ab	17.5ab

*Mean separation by Duncan's multiple range test, 5% level.
PEG, polyethylene glycol; BA, benzyladenine; GA, gibberellins.
Source: Tanne and Cantliffe (1989).

to warm, clay soils also are less suitable for early season production. Although fine-textured soils may require greater cultural management, their typically high native nutrient content can be an advantage that justifies their use.

Well-drained organic soils are highly prized for the production of carrots, parsnips and celery. These soils permit excellent early seedbed preparation as well as rapid establishment of celery or other transplanted crops. Their high organic matter content provides very good moisture-holding capacity, and through mineralization they can contribute fair amounts of nitrogen for crop growth. However, peat soils are subject to loss from oxidation and wind erosion, especially when fields are fallow or before a crop canopy is established. Strong winds occasionally uncover recently planted and even well-established seedlings. Companion crops such as oats or barley sometimes are planted on muck soils as barriers to assist establishment of slower-growing plants. When the vegetable crop is established, the companion crop is killed by herbicides, or removed with cultivation.

Celery producers favour organic soils for the ease with which they can be moved up against celery petioles to cause blanching, a labour-intensive practice that has largely disappeared, but still is an important practice for some markets. Blanching involves covering the celery petioles, but not the leaf blades, with soil or other materials that exclude light.

In newly reclaimed peat soils some producers avoid initial cultivation of celery and other vegetables. Instead one or more crops of a cereal grain precedes vegetable plantings to help in the initial decomposition of fresh organic matter, improve bulk density and possibly improve the mineral nutrient content (Guzman *et al.*, 1973).

Soil salinity inhibits seed germination and interferes with crop growth by restricting moisture and nutrient acquisition. However, when properly managed, some saline soils are capable of high yields. Most vegetable umbellifers are moderately tolerant of saline conditions.

Soil physical characteristics significantly influence crop choices. For carrot, parsnip and other root crops it is important that soils are friable and deep so as not to interfere with storage root length, enlargement and shape, an advantage organic soils readily provide. In mineral soils, deep tillage can loosen soil and disrupt compaction that might limit growth.

Soil reaction (pH)

The importance of soil pH is its influence upon nutrient availability. For most umbellifers, the preferred pH for many mineral soils is about 6.5, and about 5.8 for organic soils. Carrots tolerate a relatively broad range of soil pH, so that satisfactory growth can occur in soils with pH levels ranging from 5.0 to 8.0. However, in organic soils a pH level between 5.5 and 6.5, and between 6.0 to 6.8 in mineral soils is preferable. Nevertheless, in some areas of the UK, the USA, Canada and elsewhere carrot crops are successfully grown in some soils with pH levels well above neutrality.

Soil moisture

The availability of appropriate moisture throughout the growing season is one of the most important production requirements. Proper maintenance of soil moisture status is very important. Soil moisture requirements of the vegetable umbellifers vary considerably. Celery with a relatively shallow root system requires a frequent and large supply of soil water. Other umbellifers have less demanding moisture requirements, but none can be considered drought resistant. Crops such as water dropwort are grown in semi-aquatic conditions.

Low moisture stress early in the season will delay growth and reduce yield. Moisture deficit later in growth may decrease quality. Foliage vegetables are sensitive to low moisture at all growth stages. Those grown as seed crops are most sensitive during fruit set and maturation, and root crops are most sensitive during rapid storage root enlargement. Storage root crops like carrots, parsnips and arracacha are intolerant of waterlogging.

Excessive soil moisture saturates soil pore spaces and thus limits oxygen. This condition will limit nutrient and water absorption, increase the process of nitrification in the soil, and is favourable for attack by soil pathogens. Periods of excessive soil moisture during growth will decrease carrot root colour, length and shape, and increase the number and size of fine fibrous roots (Bradley *et al.*, 1967). The incidence of carrot root splitting is increased with a rapid change in soil moisture content as brought on by rain or irrigation, especially near the harvest period. Splitting is often prevalent when moisture is suddenly available after a period of inadequate supply. A uniform and adequate moisture supply during growth is important for achieving desired root conformation and surface smoothness of root crops.

Soil moisture should be near field capacity throughout growth. A desirable level of water-holding capacity in many soils is about 125 mm m^{-1} of soil depth with the soil water table at least 75–90 cm below the surface so as not to inhibit soil aeration. Where drainage is inadequate, it can be improved with ditches or sub-surface drains. Sub-surface systems are preferred since they interfere less with cultural operations than open ditch drains.

Proper seedbed moisture is very important for small-seeded crops such as carrot, celery and essentially most umbellifers in order to obtain uniform seed germination and seedling emergence, as well as for rapid seedling establishment after transplanting.

PRODUCTION PRACTICES

Tillage and seedbed preparation

Tillage has several functions that include incorporation of plant debris, controlling weeds, decreasing soil compaction, and improving soil aeration and moisture status. Depending on field conditions, soil preparation may involve tillage practices such as: deep chiselling to reduce plough pans or impervious layers, mouldboard or disc ploughing, discing, surface smoothing and levelling. Fall ploughing can be beneficial in areas where freezing occurs as this tends to reduce clod size and enhance soil aggregation. Deep ploughing to depths of 40–50 cm may be necessary, especially for root umbellifers, to eliminate sub-surface compaction. After incorporation of the refuse of a preceding crop, sufficient time should be allowed for the thorough decomposition of the organic matter. The presence of crop refuse before planting, especially for root crop production, may interfere with root growth, shape and smoothness.

Discing, roto-tillage or harrowing followed by floats, rollers or drags are used to further reduce clod size in preparing an acceptable surface tilth and to firm seedbed surfaces. However, excessive tillage to remove all clods is unnecessary because that increases production costs, adversely affects soil texture and may increase the tendency for soil crusting. The application of soil conditioners to stabilize seedbed surfaces generally improves seedling emergence in situations where heavy rain follows sowing, or where soil crusting occurs. Phosphoric acid (H_3PO_4) is one of several materials that can be used as an anti-crusting agent.

Well-prepared seedbeds, whether raised or flat, improve uniformity of sowing depth, subsequent emergence and seedling development. Raised beds improve drainage and soil aeration, and facilitate furrow irrigation, cultivation and other cultural operations. Raised beds in cooler climates can also contribute to slightly increased soil temperatures, which in some situations can reduce the growing period by several days to as much as 1–2 weeks. Uniform flat field surfaces and/or uniformly shaped raised beds are critical for machine sowing in order to obtain effective seed placement, good seed-to-soil contact and uniform emergence.

Soil preparation practices other than conventional tillage include no-till or minimum tillage. These procedures reduce some preparation effort and cost, and generally are less damaging to soil structure. However, minimum tillage use is dependent on good weed controls especially since umbellifer seedlings are poor competitors. Strip tillage is a version of reduced tillage where only the narrow strip of soil in which the seeds or transplants are placed receives tillage. Other practices occasionally employed are living mulches, where crop residues are finely chopped but left on the surface to smother emerging weeds and to conserve moisture. The mulch is moved away from the area where seed or transplants are placed. Benefits of no-till or minimum tillage often are not offset by the inconvenience of the practice, and therefore have not been widely used for most vegetable umbellifers. Inter-cropping is infrequently practised in intensive vegetable production, but when employed, crops selected should be compatible.

Fumigation

Soil fumigation is commonly practised for carrot and to a lesser extent for other umbellifer root crop production, mostly to control nematodes, although some disease and other pest control benefits are obtained. However, it is

Seed priming

Practices such as seed chitting and wet priming are used to improve seed field performance. However, obvious disadvantages are the need to sow wet or moist seed without significant delay and avoiding seed damage during handling and planting. Procedures for handling and mechanical planting of pregerminated or primed seed have undergone considerable investigation.

Uneven seed germination is a characteristic of most umbellifer plants because of the variable seed maturity that comes with an extended development of floral development. Seed priming is a procedure employed to overcome this problem. When sown, primed treated seeds tend to germinate rapidly and uniformly. Other advantages of priming are improved performance in stressful field conditions, such as cold or saline soils. Additionally, priming reduces high temperature-induced dormancy, such as occurs with celery.

A disadvantage of priming is that seeds are subject to accelerated ageing. Additionally, the process is time-consuming, expensive and can be technically demanding. Accordingly, the procedure is generally performed for, rather than by, the grower. The protocol for priming treatments varies for different species and even cultivars with regard to osmoticants, concentration levels, temperatures and treatment time. An additional variable is the drying procedure. In general, most priming procedures confine the seeds with an osmoticant material in a liquid or solid matrix. Adjustments in the procedure are made to accommodate the quality and condition of the candidate seeds. In some cases, priming presents an opportunity for additional treatments (e.g. growth regulators) to further enhance seed performance. Some of the specific technologies of priming and performance enhancement are proprietary.

Seed size

A widely accepted view of many producers is that large seeds are more productive than small seeds. However, the attribute of size alone can be misleading without further knowledge of seed maturity and other qualities. There are many indications that large seeds because they contain large endosperm reserves tend to produce larger seedlings with early and higher yields, although yield differences often tend to diminish when growth periods are extended. On the other hand, excellent performance can also occur with smaller seeds.

Benjamin (1982) found that variation of carrot root dry weight produced from a population of mixed-size seeds was not greater than the variation obtained with a population of graded large or small seeds. Carrot embryo development and embryo size was shown by Gray and Ward (1985) to be a more important performance factor than overall seed size or stored food reserves. They reported carrot seed weight and endosperm volume were closely and linearly related to endosperm cell number, and the relationship accounted for most of the variance in seed weight and endosperm volume. The relationship was similar for seeds of the two cultivars examined ('Chantenay' and 'Amsterdam'), and for seeds from primary and secondary umbels, although differences in cell volume were affected by season of seed production. Mechanical sizing and genetic selection have done little to reduce variability of early seedling development, and innovative methods to circumvent this problem await development.

A growing acceptance that seed size is not the sole determinant for performance has increased the practice of seed sales by count rather than by weight. Nevertheless, size grading is commonly done to improve seed drill performance.

Seed dormancy

Seed postharvest dormancy is a condition that causes erratic seed germination and variation in seedling emergence and development. To circumvent this situation, celery seeds are often stored for a year or more after harvest to allow time for the natural loss of dormancy. Germination inhibitors can also be removed by leaching (Thomas *et al.*, 1975). Celery, in particular, and seeds of chervil, parsley, parsnip and carrot usually germinate at a higher percentage and more rapidly in the light at low temperatures than in the dark at high temperatures. Celery seed germination becomes thermo-inhibited during incubation in the dark, especially at high temperatures. This dormancy can be broken by light and with gibberellin treatment. Daily alternating temperatures also can alleviate or replace the light requirement (Pressman *et al.*, 1977; Tanne and Cantliffe, 1989). Response levels are cultivar-dependent as well as influenced by the position the seed occupied in the inflorescence (Thomas *et al.*, 1979). Interestingly, Thomas (1994) reported that several fennel cultivars were found to germinate better in the dark between 20 and 25°C than at the same temperatures in the light, although germination was further improved with $GA_{4/7}$ treatments.

Other seed treatments

Additional seed treatments include hot water treatment, and the use of bleach and fungicides to eradicate seed-borne diseases such as celery blight (*Septoria apiicola*), carrot blight (*Alternaria dauci*, *A. radicina*) and bacterial blight (*Xanthomonas campestris* pv. *carotae*). For protection against seedling pests seeds are also treated with insecticides. Seed treatments can enhance performance, but cannot be expected to convert poor quality seed into good quality seed. Therefore, for the best results, the highest quality seed available should be used.

Planting

Direct seeding is the least expensive method of propagation, primarily because less labour is involved, and with present-day precision drills, successful plantings are efficiently achieved. Precision planters have become a vital component of high-value carrot and other umbellifer crop production especially where uniformity of high-density populations are important. For high plant density crops like carrots, parsley and parsnip where thinning is not practised or feasible, correct stand achievement is important. To do so requires that the seed quality, seedbed conditions and planting precision are very high. The wide use of selective herbicides and effective mechanical thinning of some crops also greatly supports direct seeding practices.

A disadvantage of direct sowing is the risk of inadequate or non-uniform field populations. The risk is much greater when attempting to sow to a final stand that will not require thinning. To minimize the risk of an insufficient population, an excess of seed is sown, and the surplus seedlings thinned to establish the eventual field spacing and population. The advantage of sowing to a final stand is elimination of thinning and use of less seed. However, planting conditions and seed quality should be ideal if this practice is to be considered. Another disadvantage of direct seeding is that the crop occupies the field for a longer period, thereby requiring a longer period of cultural management.

Seed of many umbellifer crops usually are sown shallow in order to hasten emergence and to minimize soil impedance to the fragile seedlings (Finch-Savage, 1986). Carrots, parsley and celery often are sown at a depth between 3 and 5 mm, and sometimes on to the surface and without soil cover. Sowing depths greater than 10 mm are avoided for small-seeded umbellifers unless soil surface temperatures are high and surface soil moisture is low. Larger seeds, such as parsnip and sweet cicely, are planted at depths between 5 and 10 mm. Possible wind erosion that can uncover and remove seed presents a risk for direct sowing into peat soils. Additionally, high temperatures have caused injury to young seedlings growing in peat soils due to the heat absorbed by the dark soil. To remedy these situations, rapidly growing companion plants like barley or turnip are sown alongside the crop row for shade and as a wind barrier. The companion plants are later removed. Clump or plug sowings, where several seeds are closely grouped, is seldom practised for umbellifer crops because of the unfavourable effects of early plant-to-plant competition.

Sowing rate calculations are generally based on seed germination percentage, seed vigour and a 'field factor'. Field factor is a term that considers non-seed components in determining sowing rates. These include the anticipated influence of seedbed and environmental conditions, namely temperature, moisture and soil characteristics. It is used to determine how many viable seeds to plant in order to achieve the intended field population. Such calculations are especially important for crops such as carrots or parsley which are planted at very high densities and for which population adjustments are difficult to accomplish or are not feasible. Sowings usually are delayed until soil temperatures are favourable for germination. Although carrot seeds germinate over a range from 10 to 35°C, rapid emergence occurs between 20 and 30°C. For many other umbellifer vegetables, 25°C usually is close to the optimum temperature.

Seed drills

Seed drill performance is critical for sowing effectiveness of vegetable umbellifers because these machines affect uniformity of spacing, seed depth, seed-to-soil contact and soil cover. Such machines range from simple random seed distributors to those that rapidly and precisely sow individual seeds (Fig. 5.2). Seed-coating technologies that transform small and irregularly shaped seeds into pellets of uniform size and shape have greatly improved planting practices and seed drill effectiveness. In addition to the appropriate planting equipment, a well-prepared seedbed and adequate soil moisture enhance the probability for successful emergence.

Fig. 5.2. Six bed carrot planter sowing two bands of multiple seed lines per bed. The equipment shapes each raised bed immediately before seed are sown and lightly covered with soil.

Fluid drilling

An alternative seed handling and sowing practice is to sow pre-germinated seed. Researchers, particularly at the former National Vegetable Research Station (NVRS), now Horticulture Research International (HRI) in Wellesbourne, UK extensively investigated pre-germinated seed-sowing practices. Their major objective was to improve emergence reliability, especially with regard to variation in seedbed moisture at the time of sowing in order to obtain early and uniform seedling emergence within a narrow time span after sowing. This was approached by first providing a controllable environment favourable for seed to begin germination. At or slightly before the appearance of radicle protrusion the seed are mixed with a gel medium. The gel containing the suspended seed is pumped from a specially designed planter and extruded into the soil as a ribbon-like fluid. That research resulted in a practice which became a commercial reality known as fluid drilling. The gel medium serves to transport the seeds, protects the injury-susceptible seeds and provides some residual moisture to support the emerging seedlings after sowing. However, the inability to easily separate seed for accurate spacing, the short holding ability of germinating seed to accommodate planting delays, and the requirement for specialized handling and fluid-drilling equipment has limited wider utilization of the procedure.

Transplanting

Transplanting involves the propagation, associated handling and field establishment of seedling plants and usually is the propagation method of choice for celery and several other umbellifer vegetables. Benefits include: earlier production when field conditions prevent early soil preparation and direct sowing; full populations of uniformly spaced plants; elimination of thinning; and less total seed use. Additionally, the field occupancy of the crop is reduced which may permit production of more than one crop per season on the same land. Other benefits include a reduction of fertilizer and irrigation inputs because the period of field growth is reduced, and more opportunities for weed removal before transplanting. Furthermore, there is the widely held view that transplants usually improve crop uniformity and yields. Major disadvantages are the higher investment of labour and capital compared to direct-seeding practices.

The two transplant-raising procedures are open field and sheltered nursery production. The traditional method of open field production of transplants for subsequent field vegetable production has declined considerably. However, such production continues to occur, especially in regions where field conditions and labour availability permit. Transplants are raised in specially prepared seed beds near the field or in soil in protected plant beds. After a suitable stage of growth is attained the plants are lifted as 'bare-root' seedlings for transplanting. Grower control of growing conditions is less precise and a high level of plant variability is common compared to plants produced in most sheltered nurseries. Because it was a less costly and more simple method compared with sheltered nursery productions open field transplant production was more common. However, current systems in sheltered facilities, although not simple, often are able to produce seedling plants at a lower per unit cost than field production. Even where the cost is not less, a higher level of uniformity and quality generally justifies additional production costs.

Field-grown transplants

If diseases, pests or weeds present a threat to plant growth, seedbed preparation for field-grown transplants may include fumigation. Also, if necessary, seeds can be hot water-treated before sowing in order to eradicate seed-borne diseases such as *Septoria* and *Xanthamonas*. Stored rather than fresh celery seed is commonly used to avoid possible seed dormancy. To facilitate precision mechanical sowing, seeds are size-graded and coated as uniform pellets.

To obtain high plant populations celery seeds are sown in closely spaced rows on relatively wide beds; raised beds are preferred. Rows are commonly planted 10–12 cm apart with 80–100 seed m^{-1} of row resulting in densities from 650 to 900 plants m^{-2}. Although seeds can be broadcast, row planting results in more sturdy plants and better uniformity. Sowing depths are shallow, seldom greater than 5 mm, and sometimes seeds are left uncovered to encourage rapid germination and emergence.

To avoid low temperature-induced premature bolting, field plantings are scheduled to avoid extended exposure to cold for susceptible plants such as celery. Row covers are used by some producers to accelerate seedling plant growth. For other situations, row covers are used for shade and/or protection against insect pests.

Sheltered facility-grown transplants

Transplant production in sheltered facilities is preferable to field production because better management of temperature and other growth factors permit production at any time of year. Other advantages are a greater ability for the physical handling and movement of plants as may be necessary during growth. In most sheltered production systems nursery-raised plants can be taken to the field with an attached root-ball whereas field-grown transplants typically are bare-rooted. The root-ball serves as a moisture reservoir to help sustain the plant during handling or holding prior to transplanting. It also minimizes root injury and improves rapid re-establishment of transplant growth.

The use of warm frames or similar structures for transplant propagation has largely disappeared because they provide limited temperature management capabilities, high labour inputs, and relatively inefficient use of irrigation and ventilation. Current transplant production facilities vary in cost

Sheltered celery production

The rationale for sheltered celery production is to accommodate consumer demand during periods when field-grown supplies are low which usually is during the coldest period of the year (winter) and in the spring before field production becomes available. There is some summer production, but most occurs during the colder periods of the year. Overall such production is fairly small, but can be locally important.

Glasshouses are preferred for the production of celery rather than polyethylene tunnels whereas plastic-clad houses are intermediate. However, many growers are capable of producing high-quality celery in tunnel culture while avoiding the high capital cost for glasshouses.

Competition from domestic as well as imported field-grown celery can be severe for producers of protected celery crops. Additionally, the fairly long storage life of celery can extend availability of field-grown produce, further enhancing competition for sheltered celery production.

In order to be profitable, growers need to be successful in efficiently producing the crop as well as its marketing. Scheduling cropping periods, location and climate, use of transplants, selection of cultivars, and seed germination are important factors in sheltered production of celery.

Coated and sometimes treated seed are used to produce seedlings for transplant production in peat blocks, or more often in multiple cell growing trays.

Once seedlings exhibit one or two true leaves, temperatures can be gradually reduced to 10°C to harden them for planting out. Further lowering of temperature during this period or soon after transplanting should be avoided as that increases the possibility of bolting. Temperatures as high as 24°C can be permitted without crop injury. High quality sturdy plants are produced at temperatures of 16–18°C. Temperatures greater than 18–20°C have a tendency to produce soft plants as does excess nitrogen fertilization.

Artificial illumination to supplement natural light can produce a better quality plant in a shorter time than would otherwise be possible. This should be incorporated into determining production schedules. The intensity and duration of supplied lighting are determined by the cost of energy. Supplemental lighting being very expensive is judiciously used. Its major benefit is to advance early growth of seedlings before they are transplanted. Transplants can be closely grouped thus maximizing the advantage of the supplied light.

CO_2 atmosphere enrichment is occasionally used, but unless light is adequate, its growth enrichment benefits are limited. Its growth-promoting benefits may be more useful and efficient with seedlings than with the established main crop.

It is important that maximum use is made of the expensive sheltered growing area. Therefore, plant densities are higher than common with field production. Paths and alleyways are minimized. For good market presentation, large stalks are preferred which are best achieved at 16 plants m^{-2}, especially when plants are grown at low temperatures and/or without heat. When good light and CO_2 are available the density can be increased to 20 plants m^{-2}.

In order to achieve high biomass, celery requires a continuous supply of water. Even a short period of inadequate water will cause noticeable wilting of celery stalks. Celery also requires an adequate and steady source of mineral nutrients to optimize growth and quality. In addition to a base fertilization applied to the soil, additional nitrogen is supplied during growth. Uniform application is effectively achieved when nitrogen is introduced with the irrigation water. A level of about 150 p.p.m. of nitrate nitrogen with each or every other irrigation produces good results. The fertilizer should be rinsed off the foliage if applied topically. Constant monitoring should reveal pest and disease problems quickly and allow for rapid correction.

Acceptable weight for marketing glasshouse-grown celery stalks usually ranges between 500 and 600 g. Occasionally, plants produced during winter and especially from minimal heat or non-heated houses or tunnels might not exceed 400–450 g. These can enter the marketing chain when the supply situation and quality are favourable.

Carrot plant spacing

Compared with most vegetables, carrot crops are grown at very high densities. In order to achieve the intended high field populations, plant rows actually consist of a relatively narrow area (band) of plants produced from randomly distributed seed or from several closely spaced parallel seed lines (Fig. 5.5). The width of the planted area usually is between 10 and 15 cm. Some drills scatter

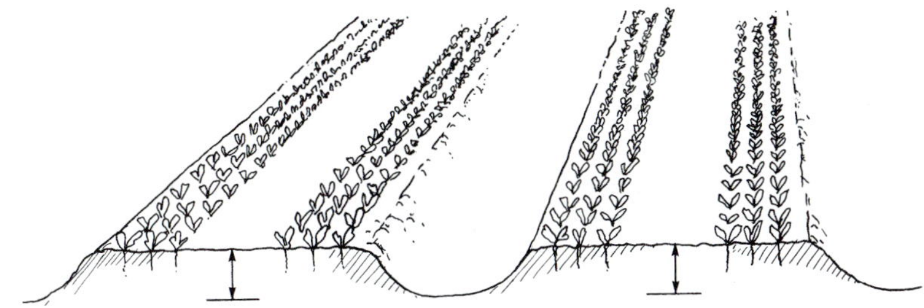

Fig. 5.5. Illustration of raised bed arrangement for multiple line high-density planting as practised in south-western USA for production of carrot, parsley and other high-density umbel crops. Distance between bed centres is about 1 m. The width of seed band commonly is about 15 cm, and bed height commonly ranges from 10 to 20 cm.

seeds fairly uniformly within this width, but more commonly three or four parallel seed lines are sown, each about 3–4 cm apart. For some production 5–12 seed lines are sown within a 15–18 cm band (Fig. 5.6). Such very high density sowings are used for the production of 'baby' carrots. These are cultivars producing small-size roots usually harvested at early storage root development, or long and slender-rooted cultivars that are processed as fresh uniform short cut and peeled segments that resemble small intact carrots.

In many carrot-growing areas, production on raised beds is the standard procedure. In the western USA seeds are sown as two bands on the bed, with each 12–15 cm band close to the edge of the bed's 50–60 cm wide level surface. The distance between these bands on the bed varies from 15 to 20 cm, and the distance between the centre point of each bed ranges from 90 to 100 cm (Fig. 5.5). Although limited, there is some use of widely spaced raised beds about 150 to 180 cm in width supporting five to six bands of sown carrots. Where level (flat) bed plantings are used the distance between seed areas (bands) usually ranges between 35 and 50 cm. Thus final plant populations are fairly similar whether flat or raised-bed culture is used. In situations where in-field winter storage is practised, the distance between rows is increased in order to enable banking soil or for placing straw or other mulching materials on to the crop.

Fig. 5.6. Example of multiple line planting shoe for high-density plantings. Seed is metered into individual tubes attached to each of the 12 openings. The width between the outermost openings is about 15 cm.

The width of the sown bands is also determined with consideration for mechanical harvester requirements. Most harvest machines for fresh market topped carrots in the USA operate by grasping and lifting the foliage and attached roots simultaneously with undercutting of the taproots. When the width of the band of plants is very wide, it is difficult for the machine to grasp the foliage effectively. Thus, many plants are not removed and harvest effectiveness is reduced. For harvesters that do not rely on top lifting, the width of the planted band is a less important factor.

Spacing adjustments can be made with some precision seed drills to correct for plant competition differences in root size due to bed edge or border effect. Having the outermost line of plants more closely spaced, and inner seed lines spaced wider, tends to equalize plant-to-plant competition, and thereby reduces root size variation.

Plant densities for carrots commonly average about 175 per m^2 for fresh market, 250 per m^2 for the production of small-rooted carrots such as 'Amsterdam Forcing' and the lightly processed fresh cut and peel cultivars, and 100 per m^2 or less for large-rooted cultivars (Lazcano *et al.*, 1998). For glasshouse production of small root carrots very high density plantings are used, and seeds may be broadcasted. Uniform sowing improves uniformity of root dimensions. A range of frequently used field populations for different carrot types and production purposes are shown in Table 5.4.

Plant spatial arrangements, between and within rows, have a pronounced influence on carrot marketable yield and quality as defined by root length and diameter. For carrots and other root crops it has been repeatedly demonstrated that increases in plant density result in a reduction in average plant diameter as well as root length. Although higher density may increase total yield, the trade-off may be that marketable yield is delayed and/or reduced.

As illustration, carrot plants of cultivar 'Long Imperator 58', spaced on a square grid for a population of 87 plants m^{-2} when compared to five higher densities produced less dry matter, but the earliest marketable yield. An equidistant population of 337 plants m^{-2} produced a greater yield, but the crop required 25 days longer to achieve market size. Further increases in plant

Table 5.4. Range of seed planting rate (millions per hectare) for different carrot types and uses.

Type	Bunching	Cello/ prepack	Bulk	Baby or cut n' peel	Slicing	Dicing	Juice
'Imperator'	1.2–2.0	1.0–1.8		2.0–3.0	1.0–1.5		1.0–1.5
'Nantes'	1.4–2.2	1.2–2.2	2.0–3.0	1.2–1.8		1.0–1.8	
'Chantenay'			0.8–1.2		0.8–1.2	0.6–1.0	0.6–1.0
'Amsterdam Forcing'		5					

density produced a greater amount of dry matter, but did not produce any marketable roots, regardless of the length of the growth period (Robinson, 1969). At very high densities, self-thinning occurs. Robinson also reported that the percentage of seedling emergence was similar over the range of 87–22,305 plants m^{-2} that he used.

Celery plant spacing

Determinations for celery plant spacing also vary according to growing conditions and market preferences and the need to be compatible with cultivation, spraying, and harvest procedures and equipment. As row space decreases, the untrimmed and trimmed stalk weight of individual plants increases. Usually plant-to-plant distances vary from 12 to 25 cm within the row, and from 30 to 75 cm between rows. Between-row spacing usually has a greater effect on yields than in-row spacing. Plant densities generally are not greater than 15 per m^2. If celery is to be blanched, wider spacing is needed to enable banking of soil against plants, or for use of other light-excluding materials that shield plants from sunlight. Blanching practices can reduce populations to as much as half of that for non-blanched celery. Alternatives to blanching are closely spaced plantings that cause a level of self-blanching or to grow light coloured (self-blanch) cultivars. While high density favours blanching, it also is more likely to be conducive for the incidence of disease and pests.

Nutrition/fertilization

Access to fertile soils with a high organic matter content and an absence of salinity or toxic elements would be expected to provide satisfactory growth, but that growth level may not be adequate for optimum yield. For intensive commercial vegetable production, growers cannot rely on soils to fully accommodate plant nutritional requirements for high yields. Accordingly, supplemental fertilizers are supplied to complement soil nutrients in order that an adequate level of nutrients are available for productive plant growth. The goal of fertilization is to assure that a balanced and readily available source of nutrients is provided throughout all phases of growth.

The major time periods, not necessarily critical, when fertilizers are applied are in advance of planting, during planting and during growth. Often the first nutrient increments are supplied from remnant fertilizer materials and the residues of preceding crops. Organic sources such as manure and compost are commonly broadcast on to the field and incorporated prior to seedbed preparation. These applications should be scheduled well enough in advance to allow for organic matter decomposition. Chemical fertilizers are also applied by broadcasting, but since nutrients are available to plants sooner they have greater flexibility as to when applications are scheduled. Typically all or most of the phosphorus and potassium component of fertilizer is supplied before planting since these minerals, being less mobile, are less subject to leaching. Because many nitrogen sources are fairly mobile and subject to leaching, only a portion of the total amount of nitrogen intended for crop production is applied initially. The remainder is supplied in one or more additions during crop growth.

An alternative to broadcasting is to apply fertilizer inserted as a band or bands close to the plant row and sometimes so that placement is beneath the seeds or plants. Application too close to the seed or plant is avoided because sensitivity to a high concentration of fertilizer salts may cause injury.

Fertilizer carried in surface, sprinkler or drip irrigation water are other side-dress applications. A practice of periodic application of crop nutrients via irrigation systems is known as 'fertigation' and has been widely adopted for vegetable umbellifer production.

With sprinkler application of fertilizers, the introduction of fertilizer is discontinued near the end of the irrigation period to wash fertilizer salts from the foliage in order to avoid tissue injury. Low concentrations of nitrogen fertilizer are sometimes applied topically during irrigation shortly before crop harvest to enhance foliage green colour. Foliage applications are also effectively used to correct some minor nutrient deficiencies.

Fertigation, especially via drip irrigation, permits very good management of the frequency and placement of fertilizers. Continuously applied nitrogen fertilizer through trickle irrigation showed a greater plant uptake of nitrogen compared to pre-plant soil-applied nitrogen. However, a high drip irrigation rate can result in excessive leaching. The relatively high soil moisture requirement of crops such as celery is unfortunately conducive to nitrogen leaching. Careful irrigation management and selection of nitrogen sources can minimize leaching losses.

Leaching of nitrate nitrogen and ammonium sulphate fertilizer is greater than losses from urea fertilizer. Yields often are higher using urea than the more readily soluble nitrate sources. However, initial growth responses usually are faster with the nitrate fertilizers.

The practice of applying a high base level of nitrogen has greatly changed because of the recognition that young plants cannot effectively use high levels of nitrogen, and that nitrogen leaching is undesirable from a production and environmental perspective. In peat soils it is common that less nitrogen is used in the initial fertilizer application. Fertilizer practices to minimize nitrate losses are recommended that consider the overall effect of previous crop residues, efficiency of crop nutrient absorption, leaching and mineralization losses.

In some soils, especially those in arid regions of the south-western USA, soil phosphorus is rapidly and strongly bound (fixed) to soil minerals so that its availability is very low. In such situations it is advisable to use freshly applied phosphorous. McPharlin *et al.* (1994) showed that carrot yield response was significantly higher to freshly applied phosphorus than to residual phosphorus except at very high rates (320 kg P ha^{-1}). In another situation, Sanchez *et al.* (1990) showed in Florida histosols that recommendations for celery phosphorus applications could be lowered because of a replacement of an earlier analysis procedure that recommended higher than necessary rates. On the other hand, for newly reclaimed peat soils, phosphorus and potassium may need to be increased twofold. Thus, soil testing is recommended to avoid excessive phosphorus application in order to minimize leaching into water systems which can lead to conditions of eutrophication.

Soils in arid regions often have a high native potassium content and thus crop response to applied potassium often is slight. However, soil tests should be made to verify that appropriate levels are available. In saline soils, fertilizers containing sodium should be avoided. In acid soils, liming may be desirable to obtain favourable soil pH levels to enhance soil nutrient availability.

Reports from numerous fertilizer practices and testing reports are quite variable, and it is difficult to make precise recommendations unless they are site-specific. Nevertheless, these reports provide information useful for a starting guideline. When precise recommendations are limited, a base level of fertilization sufficiently high to assure adequacy is usually supplied. From observed results and experience the fertilization programme can be customized for different production situations.

Unfortunately, when fertilizer is a relatively small part of total production costs, the practice of applying high levels to insure that nutrition would not be a growth-limiting factor was, and in many areas remains, fairly common. Excessive fertilization can actually diminish crop yield and quality. Excessive nitrogen nutrition results in lush growth that can increase plant susceptibility to injury and disease. Special circumstances that might require high levels of specific nutrients generally are rare.

Moreover, the movement of plant nutrients into ground water and other water bodies necessitates better nutrient management to improve fertilizer utilization and minimize nutrient, especially, nitrate loss. It is a certainty that mandated regulations for management of fertilizer volume, sources and scheduling will supplement those already in effect for pesticides and irrigation in some production areas.

Carrots

In order to accommodate the variation in nutrient availability and absorption that soils, cultural practices and environmental conditions provide, the level of fertilizers applied to carrot and other umbellifer vegetables varies considerably. For a majority of field conditions, fertilizer levels commonly used include a base application of nitrogen between 75 and 150 kg ha^{-1}; 25–125 kg ha^{-1} of phosphorus; and 0–175 kg ha^{-1} of potassium. These amounts are often complemented with an additional 75–150 kg ha^{-1} of nitrogen divided among two or three applications during growth.

Numerous fertilization studies indicate good yields are achieved at levels represented by the low end of these ranges. For many studies a typical 'best' fertilization rate is about 60 : 30 : 125 kg ha^{-1} of N : P : K, respectively. However, some of these reports often do not take into account contribution of the soil and previous crop residues and other factors that can contribute to varied results such as time of year, sources or forms of nutrients, timing, placement, time of harvest, crop quality, marketable versus total yield, stress conditions and much more. Residual effects of previous crops are important in carrot production and fertilization strategies should take into account this influence and that of rotations.

Nitrogen application before planting in soils with high organic matter often is not necessary for carrot, celery and other umbellifer vegetables. However, the presence of nitrogen in organic soils, and possibly high levels of mineralization, do not preclude supplemental nitrogen fertilizer.

In an organic soil, Hamilton and Bernier (1975) found the percentage of total nitrogen, phosphorus and potassium accumulated in the harvested portion of carrot plants from applied fertilizer was: 53.3, 51.1, and 51.8% respectively (Table 5.5). Harvested yield of the carrot crop was 49.7 t ha^{-1}, and economic yield in terms of total plant dry matter was 57.6%. Total nitrogen concentration in carrot leaves and roots were similar, 22.0 versus 21.6 mg g^{-1} dry weight, but nitrate nitrogen concentration was much greater in leaves, 3.8 versus 1.1 mg g^{-1} dry weight.

Table 5.5. Whole plant nutrient uptake of carrot and celery plants grown on organic soil and the percentage of nutrients contained in harvested portion.

	Carrot		Celery	
Nutrient	Nutrient removed (kg ha^{-1})	% Nutrient in economic yield	Nutrient removed (kg ha^{-1})	% Nutrient in economic yield
Nitrogen	213.6	53.3	159.8	49.3
Phosphorus	50.3	51.1	31.5	47.0
Potassium	273.0	51.8	318.9	49.5
Calcium	150.8	11.1	166.5	27.3
Magnesium	41.4	37.9	20.6	34.9
Nitrate nitrogen	23.1	24.7	67.9	40.0
	(g ha^{-1})		(g ha^{-1})	
Boron	260.3	45.2	139.0	43.0
Copper	72.5	48.4	34.0	43.9
Manganese	529.0	23.0	445.0	29.2
Zinc	760.0	48.6	392.0	42.6

Source: Hamilton and Bernier (1975).

Nitrogen fertilization from 0 to 146 kg ha^{-1} for carrots in mineral soils resulted in a linear increase of percentage total nitrogen with no significant difference in root nitrate or nitrite at the different fertilizer levels. It was observed that the total nitrogen accumulated, and the amount present as nitrate in the roots differed among cultivars (Chessin and Hicks, 1987). They also reported that a higher nitrogen rate (336 kg ha^{-1}) increased nitrate content, and recommended that high levels of nitrogen be avoided in carrot production intended for baby food use since the nitrate content in that product could increase. Cserni *et al.* (1989) also showed high nitrogen rates (320 kg ha^{-1}) increase carrot root NO_3 content above the permitted level for infants (400 p.p.m.). However, unless very high nitrogen rates are used, this is considered unlikely. Genetic differences in nitrate content of carrot are significant and important in cultivar development and crop production for European markets (Venter, 1979).

Carrot plants absorb proportionally more phosphorus from upper soil levels, although using phosphorus ^{32}P-tagged fertilizer, Page and Gerwitz (1969) found appreciable amounts of phosphorus was also absorbed by carrots at depths of 60 and 90 cm. Carrot yield is more responsive to freshly applied than from residual phosphorus, unless the residual source is at high levels (McPharlin *et al.*, 1994).

Phosphorus content in carrots was found to be higher when band applied, whereas potassium content was not influenced by application method, and simply increased with higher application levels. Phosphate fertilizer did not affect percentage of carrot root dry matter content compared to unfertilized treatments. In fact, differences in dry matter content were greater between years than between fertilizer levels (Evers, 1989a). Although cultivars differ somewhat with regard to phosphorus uptake, the prevailing climatic conditions during growth have the most influence on soil phosphorus availability and uptake.

Finch-Savage and Cox (1982a) reported that phosphorus fertilizer (NaH_2PO_4) applied in the gel carrier with fluid-drilled carrot seeds resulted in additional benefits of early seedling growth even when conventional pre-plant fertilization was provided. They observed that the nitrate fertilizer added to the gel with the seed reduced seedling growth and the number of emerging seedlings. Whereas nitrate might be eliminated or reduced for this procedure, the benefit of the phosphorus component could be realized. However, higher concentrations of phosphorus became phytotoxic to the seedlings.

Monitoring the availability of soil nutrients is advised in order to accommodate plant requirements for productive growth. Table 5.6 shows the range of major and minor mineral content based on tissue dry weight for productive carrot and celery growth.

Table 5.7 indicates what are determined to be sufficient nutrient levels of N, P and K for carrots crops at certain growth periods. Because of the many factors that influence its nutrient utilization, recommendations for fertilizer

Table 5.6. Fresh weight analysis of carrot leaf and petiole tissue and celery petiole tissue at different plant development periods for the range of sufficient and deficient content of nitrogen, phosphorus and potassium.

	Carrot*		Celery†		Celery‡	
Nutrient	Sufficient	Deficient	Sufficient	Deficient	Sufficient	Deficient
NO_3-N (p.p.m.)	7500	5000	7000	5000	6000	4000
PO_4-P (p.p.m.)	3000	2000	3000	2500	3000	2000
K (%)	6	4	7	4	5	3

*Carrot leaf and petiole tissue of young mature leaf at 60 days after sowing.
†Celerypetiole tissue of newest fully elongated leaf at mid-growth.
‡Celery petiole tissue of newest fully elongated leaf near crop maturity.
Source: Maynard and Hochmuth (1997).

Table 5.7. Mineral nutrient content adequacy range of carrot and celery dry matter of foliar tissues sampled at different periods of plant development.

	Carrot		Celery	
Nutrient content	Sixty days after sowing	At maturity	Six weeks after transplanting	At maturity
(%)				
Nitrogen	1.8–2.5	1.5–2.5	1.5–1.7	1.5–1.7
Phosphorus	0.2–0.4	0.18–0.4	0.3–0.6	0.3–0.6
Potassium	2.0–4.0	1.4–4.0	6.0–8.0	5.0–7.0
Calcium	2.0–3.5	1.0–1.5	1.3–2.0	1.3–2.0
Magnesium	0.2–0.5	0.4–0.5	0.3–0.6	0.3–0.6
(p.p.m.)				
Iron	30–60	20–30	20–30	20–30
Manganese	30–60	30–60	5–10	5–10
Zinc	20–60	20–60	20–40	20–40
Boron	20–40	20–40	15–25	20–40
Copper	4–10	4–10	4–6	1–3

Source: Maynard and Hochmuth (1997).

uses vary considerably for each crop and field situation. Difference between trial years often is greater than difference between treatments. The native fertility, the previous cropping history, soil type, organic matter, pH, rainfall and other production factors influence what, when and how much fertilizer is appropriate to apply. Observed crop performance and experience, especially when supported by the results of soil and tissue testing, are valuable for such determinations.

Hamilton and Bernier (1975) reported nitrate concentration for celery was greater in the petioles than in the roots, 10.3 vs. 4.9 mg g^{-1} dry weight, and that the nitrate concentration in petioles was lower than that found in leafy portions, 10.3 vs. 14.4 mg g^{-1} dry weight. Interestingly, nitrate concentration in the edible portion of celery was almost ten times that of the edible portion of carrot (10.3 vs. 1.1 mg g^{-1} dry weight).

Celery exhibits a larger response to potassium than to phosphorus fertilization, which is not surprising since celery can accumulate up to 600 kg K ha^{-1} during the growing season compared with about 40 kg P ha^{-1} over the same period (Zink, 1963; Fig. 5.8).

Parsnips appear to have a slightly lower overall fertilizer requirement than carrots. Arold (1987) found parsnip yields were similar when given 80, 160 or 240 kg N ha^{-1} and not very different from treatments receiving no nitrogen. All treatments were initially supplied with 110 kg ha^{-1} P_2O_5 and 320 kg ha^{-1}

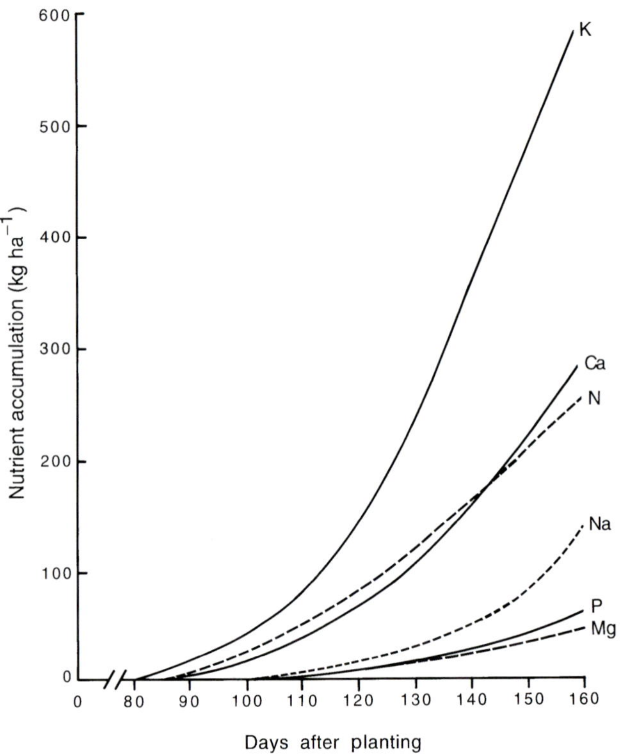

Fig. 5.8. Celery plant nutrient accumulation during a typical growth period. Source: Zink (1963).

K_2O. Application between 80 and 160 kg N ha^{-1} was optimum, although for sandy soils, the 240 kg ha^{-1} level might be more appropriate.

Essential crop nutrient functions and deficiency symptoms

Essential nutrient deficiencies are a result of inadequate levels of nutrients in the soil or environmental factors that limit their availability to plants. Conversely, nutrient toxicity is caused when excessive levels are absorbed by plants. To avoid either condition, attention must be given to soil factors such as texture, pH and moisture levels, and to maintaining the balance of these nutrients already present in soils or when supplied by fertilization. Useful information with regard to critical nutrient levels and corrective practices is available for the high volume crops such as carrots and celery and several other umbellifer crops. Although less information is available for the limited volume and lesser-known vegetables, much of the information for the major crops is transferable.

Generalities about nutrient functions and deficiency symptoms common to carrot, celery and several other umbellifer vegetables are as follows.

Nitrogen is involved in amino acid and protein synthesis, and is a component of chlorophyll. Nitrogen deficiency causes slow and restricted crop growth, roots are small, stems thin, erect and hard, and maturity is delayed. Leaves are smaller than normal, pale green, and when deficiency is severe exhibit a loss of green colour. Lower leaves are affected first. Older carrot leaves may develop a marginal red tint, whereas, leaflets of severely deficient celery become yellowish white. Parsnip leaves exhibit a vigourless spindly growth. Excessive nitrogen tends to favour succulent foliage growth, often preferentially to that for storage organs and developing seeds. Coarse-textured soils, low organic matter, low temperatures and anaerobic conditions reduce nitrogen availability. Heavy leaching of the water-soluble nitrate nitrogen can result in low and possibly deficient levels. High levels of ammonium nitrogen, especially in close proximity to roots, can cause injury.

Major roles of phosphorus are in photosynthesis, respiration, and other metabolic processes. Adequate phosphorus nutrition is associated with increased rooting and early maturity. Deficiency results in slow and stunted growth of thin, short stems and delayed maturity. Leaves of some umbellifers develop purplish coloration, typically first on the undersides. Stems and petioles also develop purple colour. Young carrot plants develop purple colour at margins of older leaves, whereas celery leaves become yellow and die early. Root and seed yield of deficient plants are reduced. Deficiency occurs more frequently in acid soils, especially when cold and wet, but high alkalinity can also decrease phosphorus availability. Excessive phosphorus uptake is seldom observed because of the fixation properties of many soils which reduces its availability to plants.

Potassium is involved in transpiration, meristematic tissue growth, sugar and starch formation, protein synthesis and also regulation of other mineral nutrient functions. Deficiency results in yield reduction with foliage symptoms typically exhibiting mottled colour or spotting accompanied by marginal leaf curl and/or burn. Older leaves develop tan or grey areas near margins and chlorotic areas may develop throughout the leaves. Beginning at leaf margins, older carrot leaves become scorched and collapse. Petioles of affected leaves become water-soaked, dry and die. Celery leaves curl back, and are shiny green with marginal and inter-veinal necrotic spots; parsnip symptoms are similar. Stems tend to be weak and root systems are poorly developed. Plants can absorb luxurious amounts with minimal effect. However, excessive potassium does increase soil salinity. Excessive leaching can increase the incidence of potassium deficiency.

The mineral elements Ca, Mg, Mn, B, S, Fe, Zn, Mo, Cu and Cl, referred to as micro or minor nutrients, are also essential for plant growth and development. Their availability in soil is closely associated with soil pH, and also in interactions with other minerals.

Calcium has a role in cell wall formation, growth and division, and nitrogen assimilation. Deficiency restricts growth, and when severe causes death of leaves, petioles and apical growth, and thus suppresses stem extension. Death of the growing point and adjacent tissues results in celery blackheart. In parsnip and carrot, deficiency causes a collapse of leaflet tissues near the petiole junction; the water-soaked tissues dry and die. Browning of core tissues in carrot root also can occur. Root growth is also restricted and root tips die. Calcium availability is low in acid and coarse-textured soils. High soil potassium and drought also limit calcium uptake, and leaching rains can further reduce calcium soil levels.

Magnesium also is essential in chlorophyll formation as well as in the formation of some amino acids and vitamins. When inadequate, older leaves exhibit deficiency as inter-veinal yellowing and chlorosis, beginning first at the margins. Leaf edges may acquire a red tint. Continued or severe deficiency causes leaf symptoms on young leaves that resemble nitrogen deficiency. Celery and parsnip leaves exhibit a similar chlorosis, although chlorosis in parsnip may begin either at margins or leaf centres. Deficiency occurs more frequently in acid soils, those with high potassium levels, and those heavily leached. Availability is also decreased in high alkaline soils. Foliar sprays with magnesium sulphate are used to alleviate deficiency.

Manganese, like Fe and Mg, is also involved in chlorophyll synthesis, and is a component of some coenzymes. When deficient, yellow mottled areas appear on the youngest carrot leaves which lack vigour and are uniformly pale yellowish green. Older celery leaves show inter-veinal chlorosis. Those of parsnip have a marginal and inter-veinal olive green chlorosis on most leaves. Deficiency occurs more often when soil pH levels are above 6.8. Most umbellifers have a moderate requirement for manganese. Manganese sulphate foliage sprays help to correct deficiency.

Boron is involved in nitrogen metabolism and water relations. When deficient, young carrot leaflet growth is greatly reduced or killed. Death of apical growth is a common deficiency symptom. Older leaves are chlorotic and curled or distorted; occasionally petiole splitting occurs. Stems are short and hard and leaves are distorted. Carrot roots are dull in colour, prone to splitting and core tissue may have hollow areas. Celery is very sensitive to boron deficiency which is exhibited as petiole twisting, distortion of young leaves and/or lateral cracks across the petioles. Axillary shoots are initiated, but fail to develop because the apical tissue dies. Newly formed celery leaves have a glossy surface, old leaves are pale and may have red margins. Celery root growth is restricted and root tips may be killed. Light brown discoloration occurs around taproot xylem and within lateral roots. Boron becomes less available in coarse-textured and alkaline soils. Celery has a higher requirement for boron than carrot. Celery cultivars vary considerably in their requirement for boron. Spray application with sodium borate can reduce deficiency symptoms.

Sulphur is an essential component of some amino acids and vitamins. Deficiency symptoms are similar to those for nitrogen. Overall growth is reduced and stems are weak. Sulphur availability is low in acid soils.

Iron is involved in chlorophyll synthesis and is a component of many enzymes. Deficiency symptoms appear as distinct yellow or white areas between veins on the youngest leaves. Iron availability is low in alkaline soils. Umbellifer crops generally have a fair tolerance to low soil Fe levels.

Zinc plays a role in chloroplast, starch and auxin formation. The first symptoms of deficiency appear as inter-venous yellowing of younger umbellifer leaves, followed by reduced shoot growth. Deficiencies are more common in acid, leached and coarse-textured soils, and those with inherently low Zn levels. High organic matter soils and high levels of phosphorus fertilization increase the possibility of deficiency. Most umbellifers have a fairly low requirement for zinc.

Molybdenum is essential in protein synthesis and some enzyme systems. Deficiency causes pale distorted development of very narrow leaves with some inter-veinal yellowing of older carrot and celery leaves. Very acid soils and high precipitation conditions are conducive to deficiency.

Copper is essential in enzyme and chlorophyll synthesis, respiration, and carbohydrate and protein metabolism. When deficient the youngest leaves of carrot are dark green and do not unfold. Older leaves appear wilted, and leaf yellowing also is common. Deficiency occurs more often in organic soils. High alkalinity diminishes copper availability. Celery tends to be susceptible to low levels of copper; carrots somewhat less so.

Chlorine is important for root and shoot growth. Although infrequent, deficiency in vegetable umbellifers results in stunted root growth, leaf bronzing, wilting, chlorotic and sometimes necrotic leaves.

Mineral nutrient requirements vary considerable between umbellifer crops. In particular, celery has a high requirement for calcium, magnesium and boron, moderate for manganese and copper, and moderate to low for zinc, and relatively low for sulphur, iron and molybdenum. Carrots have a high requirement for calcium and chlorine, moderate for copper, manganese, boron, magnesium, moderate to low for zinc, and relatively low for molybdenum, sulphur and iron. Although not an essential element, both celery and carrot usually have a relatively high uptake of sodium.

Elemental toxicity of umbellifers is unusual, although it has been reported that carrots are fairly susceptible to chlorine toxicity, and celery is susceptible to aluminum toxicity.

Irrigation

Crops benefit from an evenly distributed and adequate supply of moisture during growth and effective moisture management is essential for successful production. Crops grown for their edible foliage typically require uniform moisture throughout their development. Those grown for seeds are especially sensitive to high moisture during flowering and moisture deficits during fruit set and maturation. Moisture deficit stress during anthesis can reduce pollination and increase floral abortion.

The most critical moisture requirement for root crops occurs during storage tissue enlargement and photosynthate accumulation. Prabhaker *et al.* (1991) reported that carrot root yield, total dry matter, leaf area index, nitrogen uptake and water use efficiency increased with increasing rate of replacement irrigation from 25 to 100% of evaporative losses, and the highest yields were obtained with 100% replacement. Carrots, when grown in a wide range of soil water concentrations, usually will produce marketable roots. However, high water concentrations reduce yields more than do low water concentrations (White, 1992). White and Strandberg (1979) reported that early taproot growth in organic soils can be severely reduced by periods as short as 12 h of exposure to a water-saturated soil environment.

Pre-plant irrigation practices provide a favourable environment for umbellifer seed germination and early growth, and also improve the moisture status at lower levels of the root zone. Additional benefits are leaching of soluble salts and the opportunity to destroy germinating weed seed and early weed growth initiated by the irrigation.

Following directly sown plantings, the soil should be brought to field capacity to begin germination. Adequate soil moisture is a critical requirement during seed germination. For some crops, especially celery, even a short interruption of moisture supply can delay or abort germination. If the soil surface dries, re-application of moisture should not be delayed. Irrigation exceeding the infiltration capability of the soil is a common precursor of soil crusting. If crusting occurs, irrigation, preferably with sprinkling applied for short periods on several consecutive days is useful for softening crusts. With surface irrigation this may require maintaining water in the furrows to keep the soil damp until the seedlings have emerged. The latter practice is more likely to result in soil waterlogging which is undesirable and should be minimized as much as possible.

Transplanted seedlings should be irrigated as soon as possible after transplanting, preferably by overhead sprinklers. Transplants often are given a thorough wetting before field placement to alleviate possible desiccation. When furrow irrigation is used, the soil in the immediate area of the transplants should be brought to field capacity.

A frequent practice for some south-western US carrot producers is to use portable sprinkler irrigation to germinate the crop. Following seedling establishment the sprinkler system is removed and further irrigation reverts to furrow applications. Other producers leave the irrigation system in the field ('solid-set') throughout the growing period up to harvest. The investment in irrigation equipment is partly compensated for by labour reduction. With all sprinkler systems, of which there are many kinds, it is preferable that discharging nozzles be relatively close to the surface to improve uniformity, minimize drift and lessen water impact damage to small plants and to the soil surface.

The quantity and frequency of irrigation are influenced by stage of crop development and seasonal environment. During the autumn to spring carrot production period in the desert valleys of the south-western USA about 60–90 cm of water are generally required to complete a crop. This usually is applied in 10–12 irrigations at 7–10-day intervals. In some other carrot production areas as little as 40 cm of water distributed in four to five applications is sufficient, and for some production rainfall satisfies moisture needs. It is recommended for periods other than during irrigation that the water table be at least 75–90 cm below the soil surface.

Cultivation

Cultivation is generally regarded as the soil manipulation performed in the period between planting and harvest, whereas tillage refers to practices performed in preparing land for planting. The primary rationale for cultivation is to control weeds. Other benefits are improved water infiltration, soil aeration and removal of soil crusts and compaction. The mulching action of cultivation can conserve soil moisture, but can also be used to reduce soil surface moisture. Views about the extent and frequency of cultivation practices vary greatly. With some high-density crops such as carrots and especially at later growth stages, cultivation may not be possible or at best is restricted.

While some crops receive a minimum of cultivation others are probably cultivated more than is necessary, which in addition to possible root injury adds to production costs.

To be most effective, cultivation should be scheduled when soils are not wet or dry, since either condition reduces its effectiveness. In addition to mechanical hoes and other cultivation tools, a considerable amount of cultivation involves hand hoeing and hand removal of weeds. Hand cultivation at early seedling growth of some crops can be combined with plant thinning.

For most row-cropped umbellifers, small shovels or sweeps are used to destroy weeds and lightly aerate the surface soil. Cultivation should be shallow to minimize root injury. Occasionally, during carrot and parsnip growth, a narrow chisel is pulled between or near plant rows to break impervious layers, aerate and loosen the soil in order to improve root length and shape. Results are contradictory as to whether this practice, known as 'spiking', is beneficial. During cultivation of carrot, parsnip and celeriac plants, a light covering of soil may be swept towards stem bases and upper root surfaces to prevent shoulder greening.

Post-planting management

After germination, most umbellifer seedlings are relatively small, slow-growing and vulnerable to soil crusting. In addition to irrigation practices, soil conditioning materials and mulches are occasionally used to reduce the growth-restrictive influence of soil crusts.

Occasionally, post-planting mulching is performed in order to advance germination and growth when soil temperatures are low. Non-woven, porous fabric coverings, known as 'floating covers', are applied soon after drilling. Such row covers are able to advance carrot maturity from several days to as much as 2 weeks. Narrow strips of thin, usually perforated plastic film placed over the sown seed have also been used to increase soil temperatures (Finch-Savage, 1986).

It is also important to remove row cover mulches at the appropriate stage of crop growth. If removed too soon, little benefit is gained; if too late, plants become succulent and more susceptible to injury. The amount of perforation determines the ventilation capability of the covers to avoid overheating should temperatures rise. Usually these materials have pore spaces which are sufficiently small to exclude most insects, and large enough to allow penetration of water or pesticide sprays. It is important that herbicidal treatment occurs before cover placement because the covers also accelerate weed growth which will compete with crop plants.

Weed control

Weeds are common and chronic problems for crop producers because of their competition with crop plants for light, moisture and nutrients which can drastically limit yield and other crop performance. Furthermore, weeds may harbour pests and diseases, interfere with cultural and harvest operations, and can be a contaminant in some processed leafy crops or in seed crops. The primary goal of weed control programmes, chemical or otherwise, is to obtain the greatest possible reduction in weed competition without adversely affecting crop performance and the environment.

A characteristic of carrot seedlings and many other vegetable umbellifers is slow emergence, and slow early leaf growth. Consequently, they compete poorly with many weeds. The first few weeks of seedling growth are critical because plants are most vulnerable to weed competition during this period.

This critical state can be defined to be the maximum period weeds can be tolerated without affecting final crop performance. As long as crop performance is not affected the presence of weeds does not necessarily signify competition. The concept of the critical period is that some crops can tolerate weed competition for a longer period before non-recoverable loss of yield or other measure of performance occurs, whereas for others the tolerable period is much shorter.

The effect of different levels of competition on carrot crop performance was examined by Shadbolt and Holm (1956). They found 4.5 weeks after emergence, even at a weed population 15% of the weedy control treatment (220 plants m^{-2}), that storage root fresh weight, diameter, leaf area and total plant fresh weight were reduced compared with the weed-free control at the end of the season. The trend was more pronounced the longer the weeds were present, and with higher weed densities (Table 5.8). They demonstrated that the critical period occurs early within the first third of the growing season and that carrots possess the capability to recover from competition when weeds are removed early.

Harmful effects of weeds usually are directly proportional to weed density. As density increases, the damage tends to increase until a 'saturating population' is reached. Thereafter, further increases in density change crop response very little. On the other hand, when maintained free of competition for different periods early in the season, any reduction in weed competition permits the crop to approach its productive potential.

Carrot plant competition with weeds for light is hampered by slow leaf growth and low leaf area index. As early as 3.5 weeks following emergence, even at low weed populations, less than half of the available light was reaching carrot plants, and at the highest level of competition, light level was as little as 15% (Table 5.9). The results shown in Table 5.8 suggest that storage root development was retarded more than the leaves by early season competition. Competition for light possibly resulted in a larger diversion of

Table 5.8. Effect of weed competition (duration and density) on fresh weight of carrot storage root, root diameter, plant leaf area and total plant weight expressed as per cent reduction from weed-free control plants, measured at termination of competition periods (spring) and following subsequent weed-free growth until crop harvest (autumn). Plants were grown weed-free after respective competition periods.

Length of competition (weeks)	Weed stand (%)	Per cent reduction: Root fresh weight		Root diameter		Leaf area		Total plant weight	
		Spring	Autumn	Spring	Autumn	Spring	Autumn	Spring	Autumn
3.5	15	nr	11.6	12.6	3.5	8.4	3.0	28.3	+5.4
3.5	30	nr	4.5	25.3**	0.9	9.1	+7.5	38.4	+9.6
3.5	50	nr	11.6	30.1**	0.9	25.0	+5.7	52.8*	+9.1
4.5	15	73.6**	30.1**	41.2**	10.7**	40.6*	9.4	49.8**	24.1
4.5	30	81.2**	30.6**	52.7**	10.0**	49.3**	12.5	64.3**	39.7**
4.5	50	92.9**	44.5**	55.0**	17.8**	69.4**	28.7*	81.1**	41.1**
5.5	15	77.7**	38.7**	50.1**	14.0**	66.0**	26.8*	56.6**	37.7**
5.5	30	87.4**	47.4**	58.1**	23.9**	63.0**	33.7*	60.4**	28.5*
5.5	50	90.9**	61.9**	62.2**	32.5**	73.8**	36.1*	65.2**	65.5**

nr, not recorded.
*, **Indicate significantly different from weed-free control at the 5% and 1% level, respectively.
Source: Shadbolt and Holm (1956).

Table 5.9. Percentage of total light intensity transmitted to the level of carrot plants.

Weed population (%)	Time of measurement after carrot emergence (weeks)		
	3.5 (%)	4.5 (%)	5.5 (%)
15	48.8	22.9	17.3
30	36.7	18.4	14.0
50	28.2	14.9	15.2

The 100% weed population was 220 plants m^{-2}.
Source: Shadbolt and Holm (1956).

photosynthate to leaves rather than roots. It should be recognized that a different population of weed species could modify results although the principles of competition are not changed.

Depending on the specific umbellifer crop and weed species, weed management practices can vary considerably. Weed populations that are annuals, perennials, winter, summer weeds, or even volunteer crop plants will influence which management practices are most appropriate. Likewise, other characteristics of weed species such as height, leaf positioning and area will influence control management decisions. A listing of vegetable umbellifer weed pests is given in Table 5.10. These are most representative of species known as pests in North America and Europe. Obviously additional species are weed pests in different production areas.

In some situations where high field populations of perennial weeds occur such as nutsedges (*Cyperus* spp.) or bindweeds (*Convolvulus* spp.) it may be prudent to avoid growing carrots or other umbellifers until such weeds are removed. On the other hand, complete eradication to achieve pristine weed-free fields seldom is feasible or economically justified. For example, methyl bromide soil fumigation is a very effective weed control method. However, such fumigation is an expensive procedure, and often is not economically justified. Most commercial producers strive for an economical level of control rather than absolute freedom of weeds.

Well-performed hand removal and hoeing are effective weed control practices although not as efficient in terms of labour utilization when compared with other control practices. The use of manual labour is common in areas where labour costs are low and access to mechanical or chemical weed control are limited. Nevertheless, manual removal and hoeing continues to be utilized in many situations even in regions employing high production technologies.

Mechanical cultivation, sometimes in combination with pre-planting irrigation, mulches, crop rotation and solarization are also used to a varying extent for weed control in umbellifer crops.

The development of selective chemical herbicides has been an enormous benefit to vegetable umbellifer producers and is heavily relied upon. In many situations, it is the primary weed control measure with mechanical cultivation

Table 5.10. Reported weed pests of vegetable umbellifer crops.

Annual bluegrass	*Poa annua*
Barnyard grass	*Echinochloa crus-galli*
Bermuda grass*	*Cynodon dactylon*
Bindweed*	*Convolvulus arvensis,* other *C.* spp.
Black nightshade	*Solanum nigrum*
Canadian thistle*	*Cirsium arvense*
Canary grass	*Phalaris canariensis*
Carpetweed	*Mollugo verticillata*
Cheatgrass	*Bromus secalinus*
Chickweed	*Stellaria media*
Cocklebur	*Xanthium strumarium*
Common/bull thistle	*Cirsium vulgare*
Crabgrass	*Digitaria* spp.
Dock*	*Rumex* spp.
Fall panicum*	*Panicum dichotomiflorum*
Fiddleneck	*Amsinckia intermedia,* other *A.* spp.
Filaree	*Erodium* spp.
Flaxleaf fleabane	*Conyza bonariensis*
Fleabane	*Erigeron annuus*
Foxtail (green)	*Setaria viridis,* other *S.* spp.
Galinsoga	*Galinsoga parviflora,* other *G.* spp.
Goosefoot	*Chenopodium murale*
Goosegrass	*Eleusine indica*
Groundcherry	*Physalis* spp.
Groundsel	*Senecio vulgaris*
Hairy nightshade	*Solanum sarrachoides*
Hedge parsley*	*Torilis nodosa*
Henbit	*Lamium amplexicaule*
Horseweed	*Conyza canadensis*
Johnson grass*	*Sorghum halepense*
Junglegrass*	*Koeleria cristata*
Knotweed	*Polygonum* spp.
Kochia	*Kochia scoparia*
Lambsquarter	*Chenopodium album*
Leafy spurge	*Euphorbia esula*
London rocket	*Sisymbrium irio*
Lovegrass	*Eragrostis barrelieri,* other *E.* spp.
Malva (mallow)	*Malva neglecta, M. parviflora, M. pusilla*
Milkweed	*Asclepias syriaca*
Morning glory*	*Ipomoea purpurea* and other spp.
Mustard	*Brassica* spp.
Nettle*	*Urtica urens*
Pigweed (redroot)	*Amaranthus retroflexus,* other *A.* spp.
Pineapple weed	*Matricaria maticarioides*
Prickly lettuce	*Lactuca serriola*
Puncture vine*	*Tribulus terrestris*

Table 5.10. Reported weed pests of vegetable umbellifer crops.

Purple nutsedge*	*Cyperus rotundus*
Purslane	*Portulaca oleracea*
Quackgrass*	*Elytrigia repens*
Ragweed	*Ambrosia artemisiifolia*
Russian thistle	*Salsola iberica*
Ryegrass	*Lolium multiflorum*, other *L.* spp.
Shepherd's purse	*Capsella bursa-pastoris*
Sowthistle	*Sonchus oleraceus*
Spanish needles	*Bidens bipinnata*
Sunflower	*Helianthus annuus*
Wild barley	*Hordeum* spp.
Wild carrot*	*Daucus carota*
Wild oats	*Avena fatua*
Wild radish	*Raphanus raphanistrum*
Yellow nutsedge*	*Cyperus esculentus*

Weeds followed by an asterisk (*) are biennials or perennials. Note that common names for many weeds often differ between localities.

complementary to chemical control. Herbicide selectivity is relative, but this should not imply that they are harmless to the applied crop. Improper use can result in significant crop injury. Another concern deals with herbicidal residues in the current and succeeding crops. Again, the appropriate knowledge and use of herbicide products is important.

Umbellifer crop producers have a relatively extensive arsenal of selective herbicides available for carrot, celery, parsnip, parsley and fennel crops. However, herbicides used for these crops usually are not registered for use with the minor umbellifer crops, and in general approved herbicides for the production of many minor umbellifer crops simply are not available. Furthermore, the permissible use of present products is subject to change either because of expanded regulatory restrictions or discontinued manufacture because of low potential for recovery of developmental costs and high potential for liability. Because of potential effects to human health, regulation of pesticide products which include herbicides is undergoing intensive examination. The re-examination and re-registration in process or pending will present significant challenges to vegetable umbellifer producers. Development of new products continues, but testing, regulatory approval and introduction takes considerable time before their use can be recommended. Therefore, it is important to find alternatives for suitable and acceptable weed management that minimize adverse human and environmental impact.

Herbicides used in umbellifer production that inhibit meristem and seedling growth include: bensulide, trifluralin, fluazifop butyl and sethoxydim. Photosynthetic inhibitors are: linuron, prometryn and metribuzin, and products such as glyphosate and paraquat kill plants by cell membrane disruption and desiccation. New products for chemical weed control will rely on new chemistries and new modes of action.

General features regarding application and weed targets of some presently used herbicides for umbellifer crops are:

- Bensulide – soil applied, pre-plant incorporated or post-emergence, broadleaf and grasses.
- Trifluralin – soil applied, pre-plant incorporated, broadleaf and grasses.
- Fluazifop butyl – foliar applied, early post-emergence, grasses.
- Sethoxydim – foliar applied, post-emergence, grasses.
- Linuron – foliar applied, pre- and post-emergence, broadleaf and grasses.
- Prometryn – foliar applied, post-emergence, broadleaf and grasses.
- Metribuzin – foliar applied, post-emergence, broadleaf weeds.
- Glyphosate and paraquat are useful non-selective foliar-applied herbicides.

These examples illustrate some products and uses. It is important that all product label information is complied with before use.

The expansion of biological approaches for controlling weeds is rapidly advancing. Transgenic introduction of herbicide resistance has been achieved with corn, soybean, cotton and canola. Although umbellifer crops have not yet benefited from that approach for weed control, the potential for additional weed-combating weapons is encouraging. Broadened exploitation of diseases and insects that attack weeds are other avenues of biological control. As new technologies are introduced, the possibilities for management alternatives increases.

Crop rotations

Crop rotations are an important management practice to limit disease, nematodes, insect pests and weed infestations. Soil maintenance and its improvement are other benefits of rotations, especially if selected crops can be included that return substantial amounts of biomass and nutrients to the field. Generally, umbellifers should not follow other umbellifers, and root crops should not follow root crops. There are exceptions where carrots and other crops have been grown repeatedly on the same soil for many consecutive years. However, such situations are dependent on soil fumigation practices which are expensive and not always desirable.

6

Diseases, Disorders, Insects and Other Pests

INTRODUCTION

A substantial number of diseases, disorders and pests are known to affect umbellifer vegetable production. To effectively avoid and/or control these crop-damaging organisms and conditions, it is essential to understand the biology and functioning of the inciting entities, and the conditions that favour their development. For management strategies to be effective they must provide acceptable levels of control, be feasibly accomplished, environmentally safe, and cost-effective. Prevention and control management relies on a wide array of chemical, biological and cultural methods which, when well-integrated and compatible with other production practices, provide effective management.

PHYTOPLASMS

Phytoplasms (also called mycoplasms) are primitive plant disease organisms that lack an organized nucleus and cell wall. They are very small and able to pass through bacterial filters. Phytoplasma are found in phloem tissues and can be transmitted by grafting and by insect feeding, but not mechanically.

Aster yellows affects a wide range of plants that includes carrot, celery and several other umbellifers. This phytoplasma, of which several strains are known, is efficiently transmitted by the aster or six-spotted leafhopper *Macrosteles fascifrons*, which feeds on a very wide range of host plants and can travel, via air currents, long distances during the growing season. *Aphrodes bicinctus* and other leafhoppers also are vectors. Aster yellows is widely distributed and causes severe yield and quality reduction of infected plants.

The first disease symptoms in carrot and celery appear about 10 or more days after leafhopper transmission. Leaves become stunted, twisted and light green to yellow at the stem apex and leaf bases. A profuse mass of short,

Abandoned fields, growth of volunteer carrots left after harvest, nearby host crops and weeds can serve as a source of inoculum (Falk, 1994). Management tactics include avoiding production near already-infected carrot fields and other umbellifer crops, removing volunteer carrots and host weeds, and avoiding or controlling aphid populations. Insecticide application to kill overwintering aphids is thought to reduce the spread of virus to early spring plantings. Effective genetic resistance has been developed but has not been routinely bred into new cultivars.

Carrot Thin-leaf Virus (CTLV), in addition to its role in motley dwarf diseases, is individually an important disease. It is transmitted by willow aphids as well as the green peach aphid, *Myzus persicae*. Symptoms appear as very thin leaflets that sometimes are distorted and show a mosaic pattern. The virus survives in infected host plants, usually volunteer carrot plants. A different virus (Parsnip Mottle Virus) affects parsnips as well as celery and carrots. Its symptoms are similar to CMDV, but without the leaf reddening.

Celery Mosaic Virus (CeMV) is also known as Western Celery Mosaic Virus and Crinkle Leaf Virus. Several strains differing in virulence have been identified. Most vegetable umbellifers grown in North and South America and Europe can be affected, although symptoms are most extreme in celery (Walkey and Cooper, 1971). Disease causes foliage yellowing, stunting (especially of the central petioles) and an almost horizontal growth of outer petioles that make the plant look flattened. Symptoms appear first on the youngest leaves as vein clearing, leaflet puckering, and inter-veinal mottling. Older leaves are speckled with green bands. Brown spots on upper surfaces of outer leaves enlarge and coalesce into streaks along leaf margins. When disease is severe, leaflets are narrow and twisted, and petioles exhibit long white streaks. A crinkle-leaf strain of the virus produces extreme leaf distortion.

Cultivated plants and weeds are hosts for the virus and several aphid species are vectors. Aphids rapidly acquire and transmit the virus, but lose infectivity within a day. Aphid control is usually ineffective because of the frequent influx of new populations. Control is achieved in California production regions by the imposition of 'celery-free periods'. This procedure, during which no celery or other umbellifer crops are present, takes advantage of the short virus retention time and therefore rapid loss of infectivity by aphid vectors. Thus, when production resumes after the celery-free period, the resident aphid population will have lost infectivity. Concurrent control of CeMV weed hosts is an important part of this control programme to prevent aphids acquiring the virus. Roguing of suspect and infected plants is also important.

Cucumber Mosaic Virus (CMV) is distributed worldwide and affects many plant families. There are many strains, one is known as Southern Celery Mosaic Virus, another as Celery Calico Virus. Among umbellifers, celery is most affected by CMV. Young infected leaves exhibit outward and downward curling resulting in an open and flattened appearance. Plants are stunted, leaflet veins become greenish yellow and leaf blades are mottled, with green areas thicker than the yellow. Brown, sunken and translucent spots of irregular shape and size may develop on petioles which causes them to become shrivelled.

Primary virus sources are weed hosts and infected crops. Many species of aphids are vectors. Aphids require no latent acquisition period, but have a very short period of infectivity. Control relies on limiting the initial supply of inoculum, the use of disease-free transplants, eradication of weed hosts and destruction of crop refuse soon after harvest. Insecticides are usually ineffective because of the continuous influx of new aphid populations.

OTHER CELERY VIRUSES

Celery is affected by more viruses than are other umbellifers. In addition to the more prevalent CeMV, and CMV, several other viruses known to cause economic crop injury to celery and other umbellifers include: **Celery Latent Virus** which is symptomless on celery and nearly without symptoms on celeriac except for light green leaf spots. The virus is not aphid-transmitted, but is seed-borne and mechanically transmitted. It is reported to cause significant yield losses of celery in Europe. Control involves the use of virus-free seed. **Celery Spotted Wilt Virus**, caused by a strain of Spotted Wilt Virus, produces many small, yellow spots on older leaves. Basal areas of central petioles turn brown, and other areas may become necrotic. Early infection results in extreme stunting of plants. The virus is transmitted by thrips. Because the virus persists in plant hosts, disease management relies on controlling thrip populations. Insecticides often do not provide adequate control because new populations of thrips continue to invade crop fields. **Celery Yellow Vein Virus** is caused by a strain of Tomato Black-ring Virus which has a wide host range, and perhaps the involvement of an additional unidentified factor. The main symptom is reduced plant vigour. The virus is seed- and soil-borne and vectored by nematodes (*Longidorus* spp.). Control measures are the use of clean seed, soil fumigation for nematode control and rotations with non-host crops. **Strap-leaf Virus**, caused by Strawberry Latent Ringspot Virus, produces stunted, crinkled and strap-shaped leaves and distorted petioles. The virus is seed-borne, transmitted mechanically and vectored by the dagger nematode (*Xiphinema diversicaudatum*). Control relies on virus-free seed and transplants, and nematicides. **Celery Yellow Spot Virus** injury appears as irregular light green to yellow spots or discolored strips along veins and on leaflets. Infected umbellifers serve as primary source of this virus. **Celery Yellow Net Virus** damage is restricted to celery, parsley and carrot. Plants are dwarfed and leaves misshapen. Other symptoms include yellow flecking on leaves and dead, streaked areas on petioles that expand and kill outer leaves. The virus appears not to be soil-borne, but can be mechanically transferred, and vectors are unknown.

BACTERIA

Bacteria are microscopic, unicellular, prokaryotic organisms with a distinct cell wall. They exist as saprophytes, parasites, symbionts or pathogens, and are disseminated by mechanical means, insects, nematodes, seed, wind, splashing and flowing water. They enter plants through natural openings or wounds. Bacteria which produce pectin-dissolving enzymes incite soft rot diseases. Diseases exhibiting wilting are the result of bacterial masses plugging or destroying the vascular system.

Bacterial blight of celery (*Pseudomonas syringae* pv. *apii* and *Pseudomonas cichorii*), respectively known as southern and northern bacterial blight, are known to cause significant celery crop losses. Symptoms include bright yellow, small, usually circular leaf spots. Enlarged lesions become brown with a yellow halo (Fig. 6.2). Elongated rusty brown lesions occasionally occur on petioles. These bacteria persist in crop debris, host plants and infected seed. Free moisture on leaves is required for bacteria to enter stomata or wounds. Optimum infection temperatures are 20°C for *P. syringae* and 30°C for *P. cichorii*, the latter usually causing greater losses. Seed treatment, seed bed fumigation, minimized foliage wetting, good sanitation and managed plant spacing in transplant nursery and seed beds help avoid infection. Copper sprays can be beneficial. Some cultivar tolerance has been reported. Disease development in storage is retarded with low temperature.

Bacterial soft rot (mostly *Erwinia* and *Pseudomonas* spp.) can be caused

Fig. 6.2. Bacterial blight (leaf spot) of celery, *Pseudomonas syringae* pv. *apii*. Source: Steve Koike, University of California.

by one or more of the following species that attack vegetable umbellifers and other crops: *Erwinia carotovora* ssp. *carotovora*, *E. chrysanthemi* pv. *chrysanthemi*, *Pseudomonas marginalis* pv. *marginalis*, *P. marginalis* pv. *pastinacae* and *P. fluorescens*. These bacteria are widely distributed in most soils, so that inoculum is almost always present. Symptoms can appear on almost all plant parts and begin as water-soaked lesions that become brown, soft and slimy; usually with a distinct line between healthy and diseased tissue. A wet slime accompanied by a foul odour results as tissues decay. Secondary bacteria or fungi often invade soft rot lesions. Infected plants do not store well and losses are more extreme when free water is present on plant surfaces or when not rapidly cooled after harvest. The bacteria survive on plant debris in soil and infect through plant wounds. Warm, moist conditions favour infection and disease development. Rain, insects and physical transfer can spread infection. Carrot and parsnip roots exposed to waterlogged conditions, even for only a few hours, can become susceptible to infection.

Control procedures include: adequate drainage, destruction of diseased plants, rotation with non-susceptible crops and avoiding excessive nitrogen fertilization. Adequate soil potassium is correlated with reduced disease incidence. Avoiding mechanical injury to the harvested product, rapid post-harvest cooling and temperatures near 0°C during handling and storage will prolong storage life. *Erwinia* is retarded in storage by temperatures less than 5°C whereas *Pseudomonas* causes soft rot even at 0°C. Storage sanitation, chlorinated wash water, surface-dried roots entering storage and avoiding contact with free moisture can minimize losses. A low level of genetic resistance has been reported, but has not yet been incorporated into cultivars.

Carrot bacterial blight (*Xanthomonas campestris* pv. *carotae*). Carrots are the only host of this disease which occurs widely in all temperate production areas. Foliage symptoms resembling *Alternaria dauci* begin as irregularly shaped small yellow lesions that become brown and brittle. Dark brown lesions also form as streaks on petioles (Fig. 6.3). Flower stalks of infected plants are reduced in height by 20–80%, become somewhat chlorotic, are readily broken and are sticky to the touch because of the bacterial exudate. Lesions on seed stems can infect the inflorescence so that umbels remain small and flowers often do not develop. Infected carrot stecklings should be immediately rouged, because the disease will reduce seed yield and quality, and contaminate the seed.

Xanthomonas overwinters in infected plant debris, is seed-borne and is readily spread by moisture, insects or physically transferred. Moisture is required for infection and warm temperatures speed disease development (Pfleger *et al.*, 1974). Pathogen-free or hot water-treated seed at 51°C for 15 min, copper sprays, destruction of crop refuse and crop rotations help to reduce losses. Some genetic tolerance to bacterial blight has been observed among carrot cultivars and breeding stocks.

Carrot scab, previously considered an abiotic disease or the result of *X.*

Fig. 6.3. Carrot bacterial blight, *Xanthomonas campestris* pv. *carotae* on flowering carrot.

campestris infection, is caused by strains of the actinomycete, *Streptomyces scabies* (Janse, 1988). Scab is not a frequent occurrence, but is observed in Europe and North America in carrots and parsnips. Symptoms appear as dark brown–reddish lesions which develop into slightly raised pustules or as slightly sunken craters which become black and scab-like (Fig. 6.4). Punctate

Fig. 6.4. Carrot scab lesions, *Streptomyces scabies.*

lesions appear along or near ruptured and abnormally enlarged oil ducts, and where lateral rootlets emerge. The orientation of lesions may be in a vertical column, or randomly banded along the root surface, and may partly or fully encircle the roots. The organism produces conidia and persists as a saprophyte in the soil. Crop rotation and soil acidification tend to reduce the incidence.

FUNGI

Fungi are eukaryotic non-photosynthetic organisms. They exist as parasites, symbionts or saprotrophytes in unicellular, filamentous or plasmodial forms. Various kinds of spores produced from asexual or sexual reproduction serve as survival or propagule forms. Spores resulting from sexual recombination are oospores, zygospores, ascospores and basidiospores. Conidia, sporangia and zoospores are formed from asexual reproduction. These spores are disseminated by wind, rain, soil, machinery, other mechanical transfer, and by insects and seeds. Their persistence in soils and crop refuse may be short-lived or for relatively long periods.

In general, fungal diseases affecting umbellifer crops are more destructive than bacterial diseases. Some fungi are able to penetrate plant tissues directly while others require wounds or natural openings to infect the host. Varying with the organism, different stages of crop development and different plant parts are attacked (Table 6.1). Damping-off damages seedlings, whereas blights and mildews attack leaves, stems and petioles. Soft and dry rots damage roots and other tissues in the field and in storage. Fungal disease controls require an integration of chemical, biological and cultural methods. Integrated pest management (IPM) programmes have been developed for the major and some minor umbellifer crops that are economically feasible, and environmentally friendly.

FUNGAL SEEDLING DISEASES

Damping-off or plant dieback. Several fungal species are responsible for this common and widespread disease. Some of the fungi involved are: *Rhizoctonia solani, Sclerotinia sclerotiorum, Pythium violae, P. ultimum, P. irregulare, Stemphylium radicium* and several species of *Fusarium*. Damping-off affects most umbellifers as well as many other crops, and because of plant stand losses the disease probably is responsible for losses of a much greater magnitude than many other diseases. The seriousness of this disease is often not given the consideration it deserves.

In general, damping-off symptoms are early seedling wilting and subsequent plant loss from decay and death of roots and/or basal stem tissues. Another common symptom with carrots and other root crops is the death of

Table 6.1. Fungal diseases affecting umbellifer crops.

Common and scientific name	Crops affected	Symptoms	Factors influencing incidence, severity	Control
Seedling diseases				
Damping-off or die-back *Pythium* spp. *Rhizoctonia solani* *Sclerotinia sclerotiorum* *Fusarium* spp. *Stemphylium radicinum*	Most umbellifers	Reddish rusty root lesions, early death of root tips, wilting, stem base tissue decay, seedling death, excessive secondary rootlets	Soil-borne spores, cool soils (*Pythium*), warm soils (*Rhizoctonia*)	Minimize soil inoculum, fungicides, fumigation use
Major foliar and stem diseases				
Alternaria leaf blight *Alternaria dauci*	Carrot, celery, parsley	Small dark brown/black yellow-bordered lesions on leaf blades, petioles that blight and cause early foliage death	Seed-borne, spores, conidia, pycnidia, warm temperature, high humidity, prolonged leaf wetness	Clean or treated seed, fungicides, crop rotation, irrigation management, cultivar resistance
Celery late blight or Septoria spot *Septoria apiicola*	Celery, celeriac, smallage	Circular yellow leaf spots that enlarge become brown or grey with yellow halo. Elongated spots on petioles, pycnidia in infected areas	Seed-borne, spores, conidia, pycnidia in soil, wet foliage, mild-warm temperatures, succulent growth	Clean or treated seed, fungicides, avoid rank growth
Early blight or Cercospora leaf spot *Cercospora apii*	Celery, celeriac, smallage	Small, nearly circular pale yellow spots on older leaflets first that enlarge, turn brown and paper-like. Petiole lesions are elongated. Greyish mould growth	Seed-borne, spores or mycelium in soil, wet foliage, high humidity, high temperatures	Moisture management, fungicides, fumigation, crop rotation, adequate plant spacing, cultivar resistance

Cercospora leaf blight *Cercospora carotae*	Carrot	Leaf and petiole lesions resemble *A. dauci*, but are larger and more distinct. Leaf curling and when severe causes leaf death	Overwintering conidia, spores, high moisture, wet foliage	Fungicides, removal of crop debris, rotations, some cultivar tolerance
Powdery mildew *Erysiphe* spp.	Carrot, parsley, fennel, dill	Dusty white mycelium, usually on older leaves, petioles, slight leaf chlorosis, early leaf senescence	Air-borne conidia survive in soil and crop debris, high humidity, moderate to high temperatures	Resistant cultivars, fungicides
Major crown, root and storage tissue diseases				
Black crown or black rot *Alternaria radicina*	Carrot, celery, parsley	Black lesions at bases of outer petioles, advance to firm black decay of stem, and upper portion of storage root. Foliar symptoms resemble those of *A. dauci*	Spores seed-borne and persistent in soil, high humidity, wet foliage, warm temperatures	Clean or treated seed, fungicides, irrigation management, some cultivar resistance
Black rot *Stemphylium radicium*	Carrot, celery, parsley, parsnip	Invades crown tissues, can cause damping-off, leaf blight. Black scabby moist lesions with black margins anywhere on roots; mould, black conidia	Seed-borne, conidia in soil, debris, wet leaves, high humidity and temperatures	Clean or treated seed, crop rotation, fungicides
Sclerotinia soft rot/ celery pink rot *Sclerotinia sclerotiorum* *Sclerotinia minor*	Celery, carrot, celeriac, parsley, other umbellifers	Base of plant most often affected, usually at later growth stage. Infected sites have soft, watery pale brown or pink lesions, black sclerotia visible in mycelium and lesions; plants may collapse	Soil-borne sclerotia, air-borne spores, wet soils, humid conditions, cool temperatures, lush growth. Warm storage conditions	Minimize inoculum with removal or destruction of crop debris, well-drained soils, irrigation management, fungicides. In storage: prompt pre-cooling low temperatures, avoid free moisture, and use fungicides

Continued

Table 6.1. *Continued*

Common and scientific name	Crops affected	Symptoms	Factors influencing incidence, severity	Control
Crown or canker rot *Rhizoctonia carotae*	Carrot, parsnip	A band of dark brown decay occurs around the crown, and horizontal brown canker-like lesions appear mostly on crown and upper root, root forking. In storage: small hyphal knots above surface pits that become sunken brown dry lesions covered with mycelium	Sclerotia and spores persist in soil, warm to high temperatures, high humidity and soil moisture	Moisture management, raised beds, fungicides, rotation, dry soil surfaces, cultivar resistance, plant spacing, some cultivar resistance. In storage; fungicide dips, low temperature, humidity control
Crater rot of celery *Rhizoctonia solani*	Celery,	Reddish brown sunken lesions with water-soaked margins at base of stem and petioles that become sunken and dark brown; roots also affected	Sclerotia and spores persist in soil, warm to high temperatures, high humidity, high soil moisture	Moisture management, fungicides, crop rotation, some cultivar resistance. In storage: low temperatures, appropriate humidity, ventilation, fungicides
Southern blight/ Sclerotium rot *Sclerotium rolfsii*	Carrot, celery, parsley, other umbellifers	Base of leaves, top of root, and portions of root with moist and soft decay covered with white mycelium	Sclerotia and mycelium in soil, warm temperatures, high humidity, wet soils	Fungicides, fumigation, crop rotation

Phoma rot *Phoma apiicola* Other *Phoma* spp.	Celery, celeriac, fennel, parsnip	Stunted growth, leaf wilting, dark brown/black crown lesions, at or below soil line, spreads into petiole. Decay is soft, usually black resembles liquorice rot, pycnidia produces, grey mould on parsnip	Contaminated seed, persisting soil conidia, high soil moisture, cool temperatures, wounds	Well-drained soils, crop rotation, fungicides, fumigation
Cavity spot *Pythium* spp.	Carrot, parsnip	Small lens-shaped cavities below epidermis, most on lower storage root. Epidermis rupture shows cavity deeper and wider than opening	Presence of *Pythium* populations, poor drainage	Fungicides, fumigation, Avoidance of infested fields, soil testing to identify presence of *Pythium* populations
Fusarium yellows/ Fusarium dry rot *Fusarium oxysporum* f., sp. *apii* *Fusarium oxysporum*	Celery, celeriac, fennel, carrot, parsnip	Stunted growth, wilting, and leaf yellowing, vascular discoloration firm semi-dry black decay at stem, petiole bases, on and into roots, secondary roots infected, damping-off suckering in celery and fennel	Persistent spores in soil, warm soil temperatures, wet soils	Soil fumigation, soil moisture management, resistant cultivars
Grey mould rot *Botrytis cinerea*	Carrot, parsnip, celery, fennel	Soft, watery, reddish brown water-soaked lesions on root tips, crowns, leaf bases. During cool conditions may be covered with greyish mould; sclerotia present. Decay common at root tips	Persistent sclerotia and air-borne spores, mycelium in debris, high humidity, wet soils, warm temperatures	Fungicides, dry weather harvest, avoid root injuries, appropriate plant spacing. In storage: low temperatures, control relative humidity, avoid free moisture

Continued

Table 6.1. *Continued*

Common and scientific name	Crops affected	Symptoms	Factors influencing incidence, severity	Control
Parsnip canker and leaf spot *Itersonilia pastinacea*	Parsnip	Shallow lens-shaped cankers on root sides and shoulders that are rusty brown/black and associated with wounds, lateral root emergence sites, lesions deepened by decay from secondary invaders. Water-soaked irregular brown spots enlarge and destroy foliage, also infects flowers and seed	Seed-borne, mycelial growth, air-borne spores, high plant density, high soil moisture, cool to moderate temperatures	Well-drained soils, appropriate plant spacing, soil covering of crown and shoulders, fungicides, resistant cultivars, rotations. Plant schedules that avoid wet and cool soils
Phytophthora rot *Phytophthora megasperma* *Phytophthora syringae* Other *Phytophthora* spp.	Carrot, fennel, parsley	Firm, dark brown, water-soaked areas in wide bands across roots, tissues soften, white mould develops	Persistent spores in soil, wet soils, high humidity, moderate temperatures	Well-drained soils, irrigation management, fungicides. In storage: low temperatures, relative humidity not greater than 90%
Liquorice rot *Mycocentrospora acerina*	Carrot Parsnip, celery, celeriac	Charcoal black deeply sunken lesions with water-soaked brown margins on crown and petiole bases. With high humidity decay is soft and watery, spores and a dark grey mould with a light orange tint spreads to roots	Conidia in soil, high soil moisture, moderate temperature, injured tissues	Rotation, avoid injuries. In storage: low temperatures, controlled humidity, fungicides

Carrot rusty root Mostly *Pythium* spp. Several other fungi	Carrot, parsnip	Foliage wilting, marginal necrosis of older leaves. Root and rootlet lesions with surrounding rusty colour, taproot tips and rootlets die	High fungal populations, incidence higher in organic soils	Rotations, moisture management, cultivars
Violet root rot *Rhizooctonia crocorum*	Carrot, parsnip, celery, fennel, parsley	Purple leathery lesions become reddish brown, coalesce and penetrate, covered with purplish mould, imbedded sclerotia	Persistent soil sclerotia, warm temperatures, high soil moisture	Well-drained soils, crop rotation. In storage: low temperatures, controlled humidity, isolate infected roots to arrest disease spread
Rhizopus woolly soft rot *Rhizopus* spp.	Carrot	Pale brown, soft, water-soaked lesions on roots, crowns; white mycelium and black sporangia	Infected plants, mycelial growth, conidia, warm and wet conditions	In storage: rapid cooling, low temperatures, ventilation, free moisture, but avoid wilting and use fungicides

Commonly observed conditions that favour infection and fungal disease development are wet soils, high humidity, and warm or high temperatures. Prolonged foliage wetting is another factor often contributing to infection and disease spread. The most effective and primary disease control procedures are to avoid infested fields or using infected seed, and where possible to use resistant cultivars. Crop rotations usually reduce the source of inoculum. Procedures that effectively manage field moisture often prevent the incidence or limit disease development. Fungicides, when available, also serve as valuable prophylactics or disease eradicants, but their effectiveness relies on proper material selection and use which includes effective application and scheduling. For stored crops the common thread for control is to minimize conditions for disease development, namely low temperature and appropriate relative humidity. Careful handling to avoid crop injury, inspection to exclude infected crops and storage sanitation minimize possible crop loss.

the root tip which reduces taproot length and often stimulates multiple rooting. Some symptoms vary depending on the fungus involved. *Rhizoctonia* produces red-brown lesions near the crown, secondary root rot and taproot damage. *Sclerotinia* produces a pinkish colour of the decayed stem and root tissues. *Pythium* and *Fusarium* exhibit reddish brown lesions as bands that encircle rootlets. These fungi are commonly soil-borne and persist as spores or sclerotia. *Pythium* prefers cool soils, whereas *Fusarium* prefers warm soils. Crop rotations and cultural practices that minimize populations are basic control measures. Soil fumigation, flooding and fungicides are additional control methods.

MAJOR FOLIAGE FUNGAL DISEASES

Foliage diseases generally result from direct contact of a pathogen with leaf or petiole tissues, or from infected seed. Leaf symptoms can also be ascribed to the effect of diseased roots, such as with *Fusarium* yellows. On the other hand, some foliar diseases such as parsnip leaf spot progress to cause damage in the form of canker. Similarly, *S. sclerotiorum* leaf infections often progress as celery pink rot crown decay.

Alternaria leaf blight (*Alternaria dauci*) occurs worldwide wherever carrots are grown. Celery and parsley are also attacked. Symptoms appear first on older leaves as irregularly shaped small dark brown or black spots with yellow borders. These are usually restricted to leaves and resemble those of bacterial leaf blight. The lesions tend to enlarge and coalesce on leaves and petioles (Fig. 6.5a). Disease progresses very rapidly and can kill leaflets and entire leaves (Strandberg, 1983). Rain and sprinkle irrigation favour its development. The loss of foliage results in reduced root yield and can decrease the effectiveness of mechanical harvesters.

Spores survive in crop debris, weed hosts and soil. They are long-lived, especially when dry, and are spread by wind, water and mechanically (Fig. 6.5b). Inoculum can also be seed-borne. The spores are dispersed during the day. Spores germinate in the presence of leaf moisture, with infection occurring during the night. Fungal spores enter and infect through stomata. Disease incidence increases with high humidity and favourable temperatures between 14 and 35°C; the optimum is 27°C. In addition to fungicides, other control measures are: the use of disease-free or hot water- or fungicide-treated seed. Early and thorough destruction of crop refuse, crop rotations, irrigation management that avoids prolong foliage wetting, and reduced field travel help minimize inoculum spread. Some cultivar tolerance is available.

Celery late blight or Septoria spot (*Septoria apiicola*) is an important disease that attacks celery and celeriac wherever they are grown. Initial damage consists of mostly circular yellow spots on leaflets that become brown or grey with a yellow halo (Fig. 6.6). Spots on petioles are elongated, brownish

Fig. 6.5. (a) Alternaria leaf blight, *Alternaria dauci* and (b) spores. Source: Barry Pyror, University of California, Davis, USA.

and without definite margins. Outer leaves and stalks are mostly affected. These turn dark, wither, appear scorched and result in severe crop losses. Fruiting bodies (pycnidia) develop in the infected areas from which additional

Fig. 6.6. Celery late blight or Septoria spot, *Septoria apiicola*.

spores will be dispersed to cause a secondary spread of the disease. The fungus exists in debris as pycnidia, conidia or mycelium and also is commonly seed-borne. Infection occurs through stomata and by direct penetration. Heavy dews favour development and spread; mechanical spread is also common. Mild temperatures (20°C) are most conducive to spore germination. Although a cool-season disease, once infection occurs, decay is accelerated by high temperatures.

Disease-free or fungicide-treated seed reduces seed-borne *Septoria* as a source of inoculum. Extending the period of seed storage is commonly used to eliminate seed-borne inoculum because spores usually do not survive the storage period. Reduction of field inoculum levels by prompt removal or destruction of infected crop refuse and susceptible host plants is another important control practice. Seed-bed sanitation, protective fungicide application and avoidance of excessive nitrogen fertilizer are additional control measures. Rapid postharvest cooling and low storage temperatures retard disease development. Progeny from a cross of celery with the wild species *Apium chilense* have exhibited resistance to *Septoria* (Fig. 6.17a).

Early blight or Cercospora leaf blight (*Cercospora carotae*), although usually less frequent or severe than Alternaria leaf blight, nevertheless occasionally does cause significant crop loss. Cercospora infection usually occurs earlier in the growing season than Alternaria leaf blight. The disease occurs throughout most temperate regions. All plant parts may be attacked, although the roots usually are less affected. Foliage symptoms resemble those of Alternaria leaf blight, although lesions are more distinct and larger. Linear black lesions also occur on petioles. Necrotic areas near leaf margins result in leaf curling, and when severe, leaflets are killed. Grey-coloured spores are produced on the lower surface of the leaf. The fungus is not seed-borne, but conidia can overwinter in crop debris and other hosts, and are transferred by wind and water. Moderate to warm temperatures (28°C) are optimum for conidia germination. Germinated spores infect tissues through the stomata. Destruction of crop debris, clean seed, crop rotation and fungicides provide control. Some cultivar tolerance is known (Angell and Gabelman, 1968).

Cercospora leaf spot (*Cercospora apii*), also known as early blight, is responsible for considerable damage to celery and celeriac in Europe and North America. Although resembling Alternaria leaf blight of carrot, this disease is identified by small irregular to circular pale yellow spots on leaflets that rapidly enlarge, turn brown and become paper-like. With high humidity, a fine growth of greyish mould forms at the centre of the spots. Symptoms are most obvious first on older leaves. Petiole symptoms are similar, but lesions are elongated. These symptoms differ from those of celery late blight (*S. apiicola*). The absence of black pycnidia in early blight lesions is a major distinguishing feature.

Inoculum can be seed-borne, but more often is spread from crop debris as mycelium and spores. Conidia are readily spread by splashing water, air currents and other physical means. Infection occurs via natural openings (stomata). High humidity is required for spore germination and penetration. Temperatures between 25 and 30°C are optimum for disease development. Plant growth with minimal abiotic stress, raised bed culture and appropriate spacing to enhance air circulation are effective controls (Strandberg and White, 1978). Crowding in transplant nurseries or seed beds should be avoided. Additional control is provided by seed-bed fumigation, fungicides, crop rotation and cultivar resistance. A programme has been developed to forecast disease incidence in order to schedule fungicide applications more effectively.

Powdery mildew (*Erysiphe* spp.) is widely distributed in Europe, Asia and North America, and attacks carrots, parsley, fennel and dill. Disease is caused by *Erysiphe heraclei* and *Erysiphe umbelliferarum*. *E. cichoracearum* also causes powdery mildew of parsley. Symptoms appear as a dusty white mycelium on leaf surfaces that can result in early leaf senescence and reduced yield. A slight leaf chlorosis may also result, with older leaves being most affected. Air-borne spores (conidia) survive in soil and as contaminants on seed. Infection is favoured by high humidity and temperatures between 13° and 32°C. Several races of varying virulence occur. Although the disease is occasionally severe, fungicide use usually is not justified. Field selection for resistant cultivars has been successful; a single major gene in carrot confers resistance (Bonnet, 1983).

MINOR FOLIAGE FUNGAL DISEASES

Several other fungal diseases that attack leaves of vegetable umbellifers are occasionally economically damaging. These include: **downy mildew**, *Plasmopara nivea* which affects carrot, chervil, parsley, fennel, parsnip and celery, and *Acremonium apii* and *Corticium solani* that cause brown leaf spot symptoms of celery.

Several fungi cause **parsley leaf spot** symptoms. These include: *Didymaria petroselini*, *Septoria petroselini*, *Phyllosticta petroselini*, *Cercospora pastinacae* and some *Stemphylium* species. Fungi producing **parsnip leaf spot** symptoms are: *Cercospora pastinacae*, *Ramularia pastinacae*, *Cylindrosporium pastinacae*, *Phomopsis diachenii* and *Itersonilia pastinacae* which is better known for its root canker symptoms. *Nectria radicicola* causes a black rot disease of parsnip that has similar symptoms to *Alternaria radicina*.

MAJOR CROWN, ROOT AND STORAGE TISSUE DISEASES

Crown diseases most often are initiated by direct pathogen penetration, but also by pathogen movement from foliage. Conversely, infested crowns also produce foliar symptoms, and frequently progress into root tissues. Root diseases also produce symptoms reflected in above-surface tissues. Many root and crown diseases continue development during postharvest and storage.

Black crown or black rot (*Alternaria radicina*) occurs in almost all areas where carrots are grown; parsley and celery are occasionally affected. Foliage symptoms are similar to *Alternaria dauci.* Black relatively firm lesions usually first occur on lower portions of the outermost petioles, then advance to the stem, hypocotyl and upper root surfaces (Fig. 6.7a), but occasionally occur on lower areas of the root. A defined black margin separates healthy from diseased tissues. Lesion enlargement, tissue deterioration and softening is generally slow. Infection is favoured by high humidity and wet foliage. Spores are released from lesion surfaces during high temperature and humidity, and are wind disseminated (Gilbertson, 1996). Long-lived spores survive on crop debris and in soil, and can be seed-borne (Pryor *et al.*, 1994) (Fig. 6.7b). Roots may be infected directly via soil contact or subsequently from infected foliage. Tissue injury facilitates infection, but is not necessary. Lesions predispose stored roots or celery stalks to secondary invaders. As with *A. dauci*, significant foliage damage and loss reduce yield and can interfere with harvest. Control procedures are also similar to those for *A. dauci*. Because dense foliage growth prevents a thorough protective application, fungicides may not be fully effective. Fungicide effectiveness in storage also relies on achieving adequate fungicide contact. Satisfactory crop storage relies upon low temperature, good ventilation and avoiding contact with free moisture. A level of genetic resistance has been observed, but has not been incorporated into commercial cultivars (Gilbertson, 1996). A semi-selective medium has been used to assay carrot seed lots for detection of *A. radicina* and also for detection of the fungus in soil (Pryor *et al.*, 1998). Polymerase chain reaction (PCR) primers have been developed for the specific detection of *A. radicina* and *A. dauci* using DNA sequences of fragments amplified from the genome of these fungi. This will be useful for examining the genetic variability of *Alternaria* spp., and likely to further resistance breeding.

Black rot (*Stemphylium radicium*) occurs widely although the incidence is greater in warmer climates. Carrot, celery, parsnip and parsley are affected. Mainly a root disease, the pathogen can invade crown tissues, cause leaf blight, petiole cankers and seedling damping-off. Root lesions are black scabby spots that are often confused with those of *A. radicina*. Decay of infected root tips may cause root forking. Conidia persist in the soil on debris, and can be carried on the seed. Primary infection can occur through foliage or directly from soil into the roots. High temperatures (29–30°C) and very high humidity are most favourable for infection and rapid disease development. Storage carrots may become contaminated during harvest. Infection results in circular shallow lesions on the sides of roots and at the root tip. The rotted tissues are black and damp-to-wet depending on the level of storage humidity. The lesions are frequently invaded by secondary fungi. With high humidity, mycelium growth and conidia are formed. Control measures include use of clean or treated seed, fungicides, crop rotations, and when stored, clean

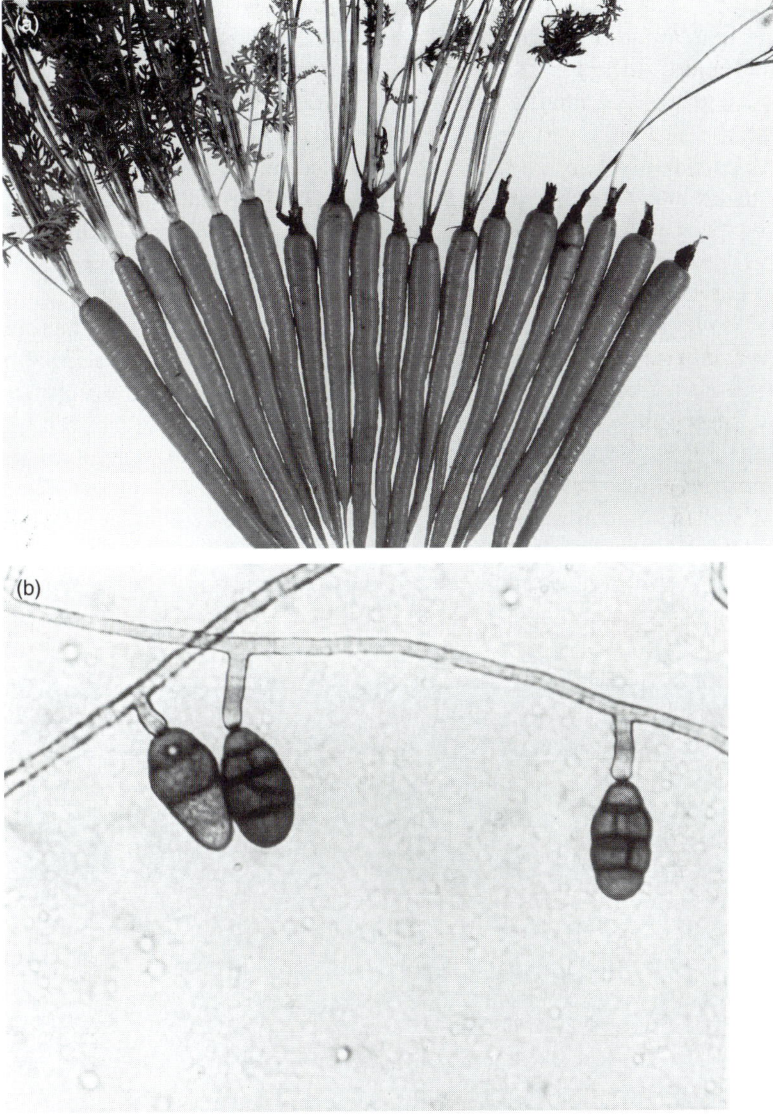

Fig. 6.7. (a) Black rot or black crown, *Alternaria radicina*. Infected roots at the right exhibit decay at the crown and loss of leaves whereas non-infected roots at the left retain their foliage. (b) Spores of *A. radicina*. Source: Barry Pyror, University of California, Davis, USA.

facilities, 0°C temperatures and relative humidity not higher than 90%; therefore, the storage period will be relatively short.

Sclerotinia soft rot/celery pink rot (*S. sclerotiorum* and *S. minor*) is widely distributed and has a wide host range that includes carrot, celery, fennel, parsley and several other umbellifers. Infection can occur at any growth stage, although symptoms most often are noticed during the late period of growth. Lesions usually occur first at the base of the plant, and when severe, the decay can cause plant collapse (Fig. 6.8). Pale brown lesions with pink-brown borders are soft and watery and frequently invaded by secondary organisms. Infected petiole bases are pinkish. Humid conditions favour mould development on infected tissues; white if *S. sclerotiorum*, buff-colored if *S. minor*. The irregularly shaped initially white sclerotia that form on and within the diseased tissues become black.

Sclerotia persist for long periods in soil. The sclerotia of *S. minor* are much smaller in size. Under favourable conditions sclerotia of *S. sclerotiorum* produce ascospores which are air-borne, but can also be moved by moisture to easily infect leaves, petiole bases and the stem area. The ascospores require wounds or senescent tissue to infect, mycelium does not. Mycelium from germinated *S. minor* sclerotia grows outward to directly infect tissues near soil level. Moisture is required for sclerotia germination and infection is favoured by cool temperatures. Pathogenic activity of *S. sclerotiorum* is optimum between 15 and 18°C; *S. minor* prefers a lower temperature. After infection the disease progresses rapidly between 18 and 25°C. Plants with incipient

Fig. 6.8. Sclerotinia soft rot of celery, *Sclerotinia sclerotiorum*. Source: Mike Davis, University of California.

infection continue to decay in storage because the fungus is able to grow even at 0°C. Of the two species, *S. sclerotiorum* is the more important storage disease.

Several cultural practices to eradicate and/or prevent increases in sclerotia populations in soil provide some control. These include: destruction and burial of crop refuse, deep ploughing, flooding, solarization, irrigation management and elimination of weed hosts, and fungicide applications provide some level of control. Fertilization that provides adequate potassium and avoids excess nitrogen, as well as rotation with non-susceptible crops, is also helpful. Crop rotation is probably the most important. Crop spacing and raised beds to improve drainage and aeration also benefit control. Some variation in cultivar susceptibility occurs, but present resistance is inadequate. Prompt postharvest cooling and storage maintenance at 0°C suppresses decay. Fungicide dips can offer additional benefits, but approved registration for fungicide use is limited. For celery, controlled atmosphere storage is useful to minimize disease advancement and to extend storage life (Reyes, 1988).

Crown rot or canker rot (*Rhizoctonia carotae*), an important crown and root disease of carrot and also parsnip in Europe and North America, is responsible for significant storage losses. Symptoms appear as a band of dark brown decay around the crown. Mycelium development is favoured by high humidity. Horizontal dark brown canker-like lesions that resemble cavity spot also appear at sites of lateral root emergence and, most often, on crown and upper portions of storage roots. Decayed tissues are fairly firm, dry, dark brown to black with clusters of irregularly shaped sclerotia (Punja, 1987). Sclerotia and spores survive in soil and attack many plants species. Control relies on raised seed beds, crop rotation, fungicides and, in some cases, cultivar resistance. Fungicidal dips before storage limit disease spread. Low temperatures and humidity control reduce further development.

Crater rot of celery (*Rhizoctonia solani*) is an important storage and field disease of celery in Europe and North America. Disease incidence tends to be more common in organic soils. The fungus is soil-borne, persisting as sclerotia and basidiospores that infect above- and below-ground tissue. Oval tan to reddish brown elongated sunken lesions appear on petiole basal tissues. Lesions have water-soaked margins that enlarge and become dark brown. Petioles can also become distorted. Field infection occurs during warm weather. At temperatures above 20°C, disease development is rapid. The use of raised beds for soil moisture management and fungicidal applications (Pieczarka, 1981) are important controls; crop rotations are also helpful. Some cultivars exhibit moderate resistance.

Mould growth on infected roots in storage is favoured by high humidity or direct moisture and occurs even at 0°C. Symptoms appear as small tufts of whitish mould on roots, and at the crown area of celery stalks. Pits beneath lesions enlarge into sunken craters. The mould often continues to spread to major effect of the disease is on the roots, although foliage is also infected. The primary symptoms are stunted plant growth and wilting due to fungus-colonized water-conducting tissues (Fig. 6.10). Leaves become yellowed and vascular discoloration can extend into the petioles. Foliage yellowing usually advances with crop development. An orange brown to black semi-dry basal decay of petiole bases and stem areas develops. When severe, the rot can result in a cavity in the interior area of the crown (root axis). Increased secondary root production, which often occurs, soon is also infected. Secondary growth of adventitious buds occasionally results in splitting of celery petiole bases and stalk deformation.

Inoculum exists as persistent spores in soil and also as a saprophyte on diseased plant debris. Spores are infectious even after many years in soil. Disease entry is usually through root tips; root exudates appear to stimulate spore germination. Low soil temperatures and cool seasonal growing conditions tend to limit symptoms. The minimum soil temperature for disease development is 15°C. Between 20 and 25°C, the disease progresses rapidly. Soil fumigation and/or sterilization provides control. Rotations are helpful, but effectiveness is incomplete because of the long viability of the spores. To minimize crop damage, it is recommended that soil compaction, low soil pH, poor drainage and movement of infected soil into fields should be avoided. Production during low to moderate soil temperature periods, and the use of resistant cultivars does permit production in infected fields (Quiros, 1997).

Other forms of *F. oxysporum* are field root diseases that often are most destructive during storage. Carrot and parsnip root symptoms occur as a

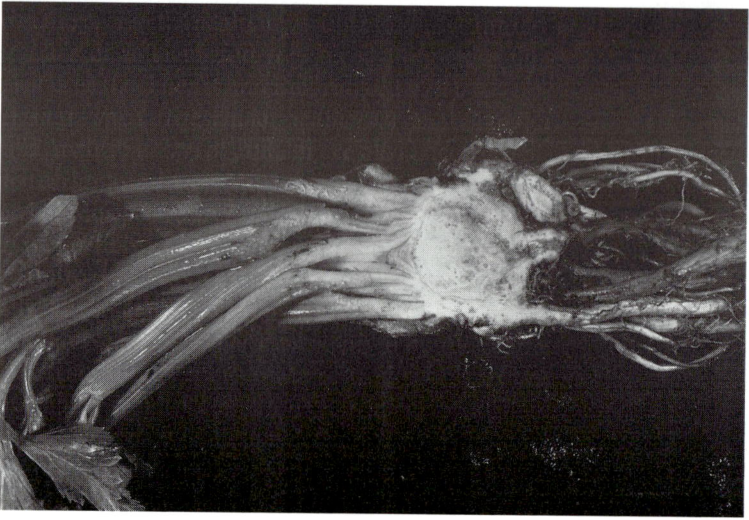

Fig. 6.10. Fusarium yellows of celery, *Fusarium oxysporum* f. sp. *apii.*

crown rot or as cankers much like crater rot. The lesions tend to be limited, leathery in texture and usually without obvious mould. Crown rot lesions are more invasive than those on root sides. Soil-borne inoculum persists as chlamydospores. Warmer temperatures rapidly advance fungus development. For control it is important to minimize crop injury and maintain low-temperature storage while avoiding free moisture on roots.

Grey mould rot (*Botrytis cinerea*) results in considerable losses for carrot, parsnip, celery and fennel producers in temperate regions of Europe, North America and Asia. Sclerotia and mycelium persist in crop debris. During mild temperatures and high humidity many spores are produced that can infect plants in the field or contaminate plants during postharvest handling and storage. Wet field or storage conditions favour infection from air-borne spores or by direct hyphae contact. The fungus generally spreads into tissues at the base of petioles or the root crown. The soft, watery, reddish brown root lesions resemble pink rot. Water-soaked tissues have a spongy appearance and become leathery and covered with mould. A white mould develops during cool and humid conditions followed by a profuse production of embedded grey-brown conidia and small black sclerotia. Sometimes symptoms appear as defined dark brown or black lesions with little mould growth. Aged tissues are more susceptible to direct infection. Crop injury and desiccation increase susceptibility. Optimum fungal growth occurs between 23 and 25°C, although infection and growth in storage occurs even at 2°C.

In the field, uniformly well-spaced plants that facilitate air circulation will reduce disease incidence. Use of approved systemic fungicides during growth, dry weather harvesting, avoiding crop damage and desiccation are suggested control practices. Rapid postharvest cooling, low temperature holding, avoiding moisture condensation on the product, a slight lowering of relative humidity, as well as careful trimming of foliage, and registered fungicides dips complement good storage practices (Lockheart and Delbridge, 1974).

Parsnip canker and leaf spot (*Itersonilia pastinacae*) is the major parsnip disease in Europe and Australia. Carrot and dill are also affected. This canker-causing fungus is also an important parsnip foliar disease (parsnip leaf spot). Cankerous lens-shaped lesions occur on root crowns or shoulders, but also are found on other root portions. These are usually associated with wounds or areas where lateral roots emerge. Affected areas are dark rusty brown or black and usually fairly shallow, but deepen when infected by secondary soft rot bacteria. Lesions resemble black rots caused by *Alternaria*, *Centrospora* or *Phoma* species. The fungus survives as resting spores in unharvested infected roots. During cool moist conditions, air-borne spores are released that fall on to the foliage causing leaf spots, others directly contact root crowns and shoulders and produce cankers. Leaf symptoms appear as small water-soaked irregular brown spots that can enlarge to destroy the foliage. The fungus also attacks flowers and can contaminate seeds (Smith, 1966). Optimum temperature for infection and growth is about 20°C.

root harvest should be considered. Infected roots should not be field-stored as mycelium readily migrates to adjacent roots. Lesions on stored roots enlarge, coalesce and deepen during storage and result in severe decay and losses. Removal and destruction of infected roots, combined with deep tillage, weed eradication, crop rotation and good drainage are useful control measures.

Rhizopus woolly soft rot (*Rhizopus* spp.). Several *Rhizopus* species, e.g. *R. oryzae* and *R. stolonifer*, are involved in this storage disease of carrot that occurs in North America, southern Europe, many Mediterranean areas and India. Its occurrence is more frequent when storage temperatures are insufficiently low. Pale brown, soft, water-soaked, often leaky lesions can be found on roots, crowns and injury sites. A coarse white mycelium and black sporangia are produced. Spores persist in soil and infection is caused by airborne sporangiospores or by hyphae contact. Decay is accelerated by high temperatures, although *Rhizopus* species differ in development rate, with temperature optima ranging between 25 and 35°C. Infection and development is poor at and below 10°C. Storage losses can be limited by rapid cooling following harvest and by temperatures maintained below 5°C, with good ventilation. Wilting, which increases susceptibility, should be avoided.

MINOR CROWN, ROOT AND STORAGE FUNGAL DISEASES

Additional fungal pests causing root and storage diseases include: **Aspergillus black mould** (*Aspergillus niger*) that attacks carrots and other umbellifers. **Acrothecium rot** (*Acrothecium carotae*), **charcoal rot** (*Macrophomina phaseolina*) and **hard rot** (*Gliocladium aureum*) which cause dry rots of carrots. **Chalaropsis rot** (*Chalaropsis thielavioides*), **Thielaviopsis rot** (*Thielaviopsis basicola*) and **blue mould rot** (*Penicillium expansium*) cause rots of carrots and parsnips. **Sour rot** (*Geotrichum candidum*) and **Mucor rot** (*Mucor hiemalis*) cause soft rots of carrot roots.

NEMATODES

Nematodes are usually microscopic, soil-inhabiting or aquatic elongated slender unsegmented roundworms that attack plants and animals. Some are parasitic, but most are saprophytes. Identification is from microscopic observation of their morphology and evaluation of molecular markers. Because of their small size, and because most feed on roots, they are not readily detected until crop injury occurs. Some nematodes feed superficially (ectoparasites) by penetrating roots, but remain in the soil while others (endoparasites) feed while partly or completely embedded in plant tissues. Nematodes are not very mobile and are disseminated by soil movement, irrigation water, wind, and mechanically, often by movement of farm equipment from infected sites.

Nematode damage of umbellifer crops is a frequent and serious threat to production worldwide. Nematode feeding and colonization of plant roots reduces vigour, and can cause severe plant stunting, wilting, chlorotic foliage and yield losses. For crops with storage roots, such as carrots, marketable yields are greatly reduced because of malformed and reduced root size, and root galls (Fig. 6.12). In addition to cosmetic damage, tissue injury also provides entry to *Pythium* and other fungal invasion. Some nematodes such as *Xiphinema heterocephalus* are virus vectors.

Root-knot nematodes, *Meloidogyne hapla, M. incognita, M. javanica, M. arenaria* and *M. chitwoodii*, are the most widespread and important nematode pests to affect umbellifer crops. *Meloidogyne* species produce root galls. Galls associate with *M. hapla* are small, whereas larger galls are produced by other *Meloidogyne* species. In addition, root forking, stubbing and plant wilting occurs. Northern root-knot nematode, *M. hapla*, is most important in Europe, Canada, the northern USA, and cool temperate regions of South America and Asia. Southern root-knot nematodes, *M. javanica, M. incognita* and *M. arenaria* are most important in warm temperate, subtropical and tropical areas. *M. chitwoodii* is a significant pest in the north-western USA.

The **cyst nematode** *Heterodera carotae* causes serious crop losses of carrots in France and other western Europe countries. Its feeding and colonization of roots of several umbellifer species greatly stunts plant growth. Additional nematodes damaging vegetable umbellifers and some of their injury-producing characteristics are:

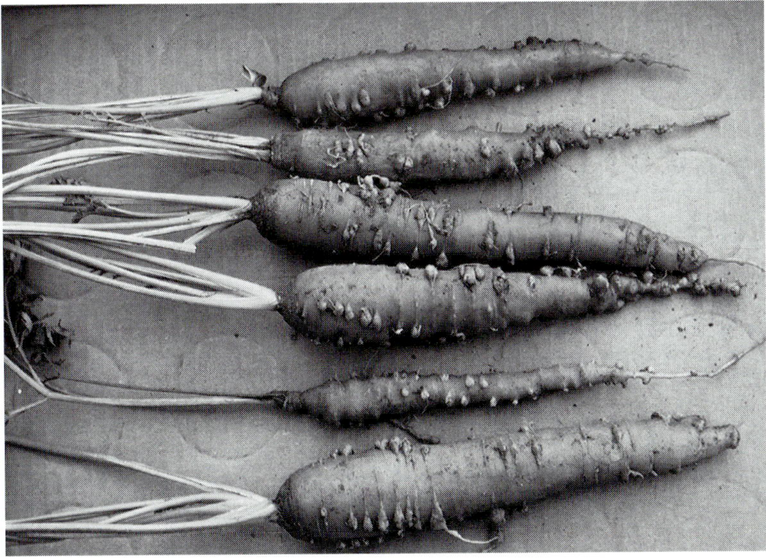

Fig. 6.12. Root-knot nematode (*Meloidogyne* spp.)-infected carrot roots.

- Sting, *Belonolaimus gracilis*: root branching and short and stubby roots.
- Pin, *Paratylenchus hamatus*: greatly limits secondary root growth and causes root necrosis.
- Lesion, *Paratylenchus penetrans*: leaf chlorosis, limited fibrous root growth and can cause root death.
- Stubby root, *Paratrichodorus* spp.: short, stubby or slightly swollen roots and death of secondary lateral roots.
- Needle, *Longidorus africanus*: cessation of root elongation, root branching, root tip swelling and tip necrosis and plant death.
- Awl, *Dolichodorus heterocephalus*: leaf yellowing and formation of many shortened secondary roots.
- Dagger, *Xiphinema diversicaudatum*: celery strap-leaf virus vector.
- Stem and bulb, *Ditylenchus dipsaci*: wounds provide entry for secondary fungi.

Fumigation practices and specific nematicides continue to be effective controls for many situations. Because nematode populations are variable, pre-plant sampling is useful to identify fields that contain populations that may require chemical treatment. Restrictions for the use of some chemical nematicides and fumigants, namely methyl bromide, has stimulated the search for alternative controls. Under development are several promising candidates for biological control and useful levels of genetic plant resistance (Roberts, 1998). Crop rotation practices also reduce nematode populations and plantings made when soil temperatures are less than 18°C can minimize crop damage.

PHYSIOLOGICAL DISORDERS

Physiological disorders, also known as abiotic diseases, often are a response to the lack or excess of essential crop nutrients or other factors unfavourable for growth. Such factors could be a lack or excess of light, moisture or temperature. Other possibilities may be poor soil structure, poor aeration, soil compaction, high soil salinity or air pollution. An often innocuous contributor to some plant disorders are chemical pesticides that also adversely affect plant development. Symptoms of these abiotic factors often are difficult to diagnose. Additional confusion about their identification occurs because some disorders mimic symptoms caused by viruses or other pathogens.

Blackheart is a significant celery disorder. Symptoms first appear on tender-growing leaflets in the central crown 'heart' area of the plant (Fig. 6.13). Tissues near leaf tips or leaf margins die. These dead areas, at first brown, become black as decay expands. Although initially a dry rot, the dead tissues are frequently invaded by soft rot bacteria and become soft and slimy. Early symptoms can be hidden behind healthy outer petioles.

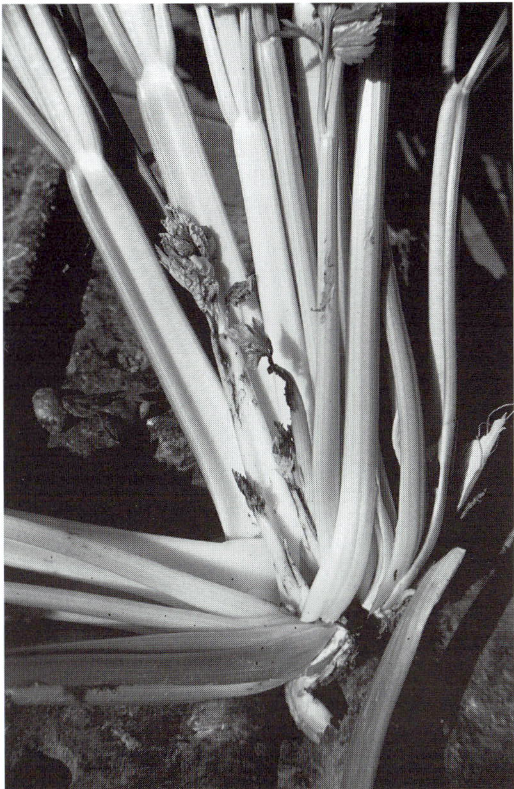

Fig. 6.13. Blackheart disorder of celery.

An interaction of related factors that include moisture stress, nutrient imbalances, low tissue calcium in particular, and conditions conducive to an excessively rapid growth rate favour the occurrence of blackheart. The incidence is more frequent in soils with a low soil pH and high soluble salt content. Control requires appropriate moisture and nutrient management especially during conditions of low soil moisture, warm temperatures and high relative humidity. Lush vegetative growth should also be avoided. The timely applications of calcium nitrate $Ca(NO_3)_2$ or calcium chloride $CaCl_2$ sprays can prevent or reduce development (Cox and Dearman, 1978). Cultivars vary somewhat in susceptibility.

Hollow heart occasionally is found in storage roots of parsnip and celeriac and is generally associated with rapid and luxurious vegetative growth.

Cracked stem and brown checking develop at right angles to the prominent ribs of the petioles and appear as a curling back of epidermal tissues with a brown discoloration of the affected tissues during celery growth, resulting in market losses (Fig. 6.14). Leaf margins and small petioles become brittle. Causative factors are deficient soil boron and other mineral nutrient interactions. High tissue content of potassium or nitrogen, especially when the ratio of ammonia is greater than that of nitrate, as well as low calcium tissue content, contribute to the incidence of this disorder. Prevention relies on nutrient management, corrective boron sprays or soil amendment, the preferential use of nitrate versus ammonium nitrogen sources, and use of resistant cultivars.

Pencil stripe of celery appears as a dark brown longitudinal striping under the epidermis extending along the petiole (Burdine, 1973). One form is characterized by strips on the outside of the ribs. Another has similar symptoms, but may be accompanied by brown checking or cracked stem symptoms resembling boron deficiency.

Rust appears as a reddish to dark brown pigmentation on inner and/or outer surfaces of celery innermost (heart) petioles. These may also manifest into pencil stripe symptoms. The incidence occurs more often after heavy rains. Suspected causes are certain fungicides, chelated nutrient sprays and a marginal boron supply, especially in the presence of ammonium nitrogen. Some cultivars are known to be more susceptible.

Other celery disorders of lesser occurrence are: **feather leaf**, **nodal cracking** and **foliage chlorosis**. With feather leaf, the secondary leaflets, usually those at the first node lose chlorophyll. This cosmetic damage usually occurs with advanced and over-mature plants. Nodal cracking describes the condition where a horizontal crack occurs just under the first node on older

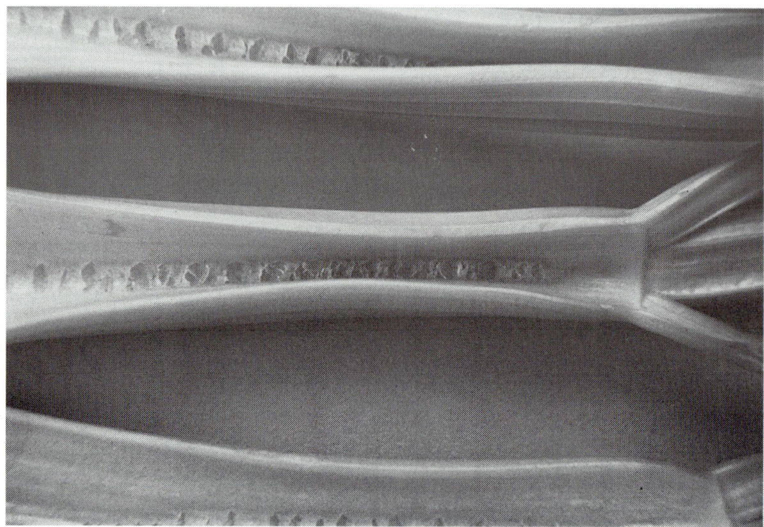

Fig. 6.14. Celery brown checking.

outer petioles. Its severity tends to be higher with high soil potassium levels and high pH soils. Foliage chlorosis can result from magnesium deficiency. Each of these disorders can be avoided or minimized by cultivar selection. Additional celery disorders are **brown stem** and **black stem**. Brown stem exhibits a dark brown, soft, watery, breakdown of parenchyma tissue. Several or all petioles can be affected. Its incidence is associated with poor soil aeration, but a definite cause is unknown. Black stem is a postharvest disorder of unknown aetiology in which dark vascular strands appear as streaks in the intact petioles. Controlled atmosphere storage is known to control the disorder.

Phenolic browning is a cosmetic disorder occasionally responsible for market losses, especially with parsnips. Root colour changes from chalk-white to yellowish brown following harvest. Blemished areas occur scattered on root surfaces. It has been proposed that the surface browning results from enzymatic oxidation of phenolic compounds during wound suberization. However, susceptibility to browning could not be explained on the basis of total phenol or oxidizable phenol content. Thus, the incidence is variable and the number of possible causative factors is unclear and unresolved. Low temperature storage helps reduce the extent of browning, and postharvest dips in 0.5% solutions of calcium chloride, ascorbic acid and/or citric acid also reduce browning. Although infrequent, browning also occurs in some processing carrots; the cause and control are unknown.

Bolting is a normal process in the life cycle of plants, and thus is not a physiological disorder. However, this is viewed as a disorder when its occurrence is premature in root, leaf and stem crops. Extended periods of low temperatures during the first year of growth of biennial umbellifers can induce bolting in the second year. Periods of high temperature and/or stressful conditions can accelerate the development of floral stems in annual plants. Premature development and elongation of the seed stalk complicates harvesting and usually reduces or eliminates the marketability of the vegetative portions of the crop (Fig. 2.11b).

Pithiness is a condition found in some tissues in which cells undergo autolysis. In celery this is characterized by a breakdown and separation of petiole parenchyma cells resulting in air spaces within the tissue (Fig. 3.9). The pithy condition can extend through the longitudinal axis of the celery petiole from the crown-attached base to the leaf nodes. Pithiness occurs more readily in outer leaves than inner leaves. Often the first leaf node is affected more than the petiole. The disorder is heritable and occurs widely in wild celery. The incidence is positively correlated with advancing physiological age, nutrient imbalance and environmental stress, especially inadequate available moisture. Water deprivation stress can induce symptoms very quickly (Aloni and Pressman, 1979). Cultivars vary in susceptibility and avoidance relies on cultural management.

Root splitting/cracking/breakage. Carrot cultivar susceptibility to these root defects differs (Dickson, 1966). In cultivars which differed in their susceptibility to splitting, splitting was shown to occur by cell wall rupture, as opposed to inter-cellular separation (McGarry, 1993). Splitting occurs when stresses within roots exceed the tensile strength of the immediate superficial tissues. The result is a radial longitudinal fracture within, and restricted to the periderm and phloem parenchyma. Splitting occurring early in root development (known as growth cracks) causes root disfigurement. The wound tissues occasionally become infected with secondary fungi. Widely spaced carrots and large roots are more likely to exhibit early splitting as well as produce more secondary roots compared with carrots that are small and/or closely spaced. Variations in growth rate and soil moisture availability are suspected causes.

Another form of longitudinal root splitting (known as shatter cracking) is initiated by mechanical damage. Some carrot cultivars are especially vulnerable. High root turgidity appears to increase susceptibility. Although water status plays a major part in determining carrot strength, differences in split susceptibility cannot be entirely explained on the basis of root turgor (McGarry, 1993).

Splitting, cracking and root breakage during harvest and subsequent handling contributes to significant losses, some of which are avoidable. For example, turgid roots are most susceptible to damage, especially when soils are cold, and such susceptibility is highest at early morning. Avoiding early morning harvest to allow some soil warming and decrease in turgor can lessen root damage. It is also recognized that the amount of cracked and broken roots incurred during mechanical harvesting and handling can be decreased by a higher level of careful operation.

Root forking (also known as fanging or sprangling) is one of the most important defects reducing marketability of umbellifer root crops (Fig. 4.6). The occurrence is often due to early damage or obstruction of the taproot apical meristem resulting in a shortened taproot often associated with the formation of branching secondary storage roots producing a distorted, multiple-rooted organ. Abiotic factors such as buried crop debris and rocky or compacted soils are implicated as contributory causes, as are root damage from freeze heaving, and high water tables. Nematodes, fungi and insect-feeding injury can also cause deformed and forked roots. Genetic variation for predisposition to root forking has been observed, but breeding for resistance is complicated, due partly to the multiple causes of this disorder.

Air pollution occasionally is responsible for plant damage. Ozone is known to be injurious to celery, parsley and carrots. High ozone concentrations can cause a bleached stippling to upper leaf surfaces of these plants. Peroxyacetyl nitrate (PAN), an ingredient of smog, injures leaf surfaces. Acid fog, with high levels of sulphur dioxide, and ethylene are other plant-injuring pollutants. Crop injury from air pollutants, while infrequent or subtle, can nevertheless have a significant effect on growth. Yield and quality are affected by early senescence, reduced photosynthesis or metabolic interference caused by the various pollutants.

Bitterness formation in carrots is caused by the formation of a bitter-tasting compound, isocoumarin (3-methyl-6-methoxy-8-hydroxy-3,4-dihydroisocoumarin). Its formation is induced by ethylene; directly or from root injury, and/or disease-evolved ethylene (Chalutz *et al.*, 1969; Lafuente *et al.*, 1991, 1996). Freshly harvested carrots, when exposed to ethylene, are more prone to develop bitterness than those stored, and young carrots become bitter more rapidly than older roots. More bitterness is found in the periderm and stem end of the root than in other portions. Fortunately carrot roots generate very little ethylene, but to avoid the disorder it is important to minimize root injuries and avoid exposure to ethylene. Increased temperatures and ethylene concentrations further increase isocoumarin formation. Refrigerated storage of carrots in the absence of ethylene sources such as apples or pears is important to avoid development of bitterness. Aster yellows infection can also cause bitterness in carrot tissues.

NUTRITIONAL DISORDERS

Nutrient deficiencies are a more common occurrence than toxicities. General symptoms of major and minor mineral nutrients deficiencies common to carrots, celery and some other umbellifer crops are discussed in Chapter 5.

PARASITIC PLANTS

Several species of parasitic plants, including dodder and broomrape, attack vegetable umbellifers. Their parasitism can severely stunt production, especially if established early in the season. Species of *Cuscuta* are climbing parasitic plants generally known as **dodder**. They occur worldwide and invade a broad range of host plants. Their occurrence in the field is marked by a network of thin herbaceous stems that have no chlorophyll. Fine branches of the orange-coloured stems grow between cells of the host plant and even finer hyphae-like stem divisions upon contacting sieve tube elements grow inter-cellularly (Fig. 6.15).

Dodder plants obtain their nutrient needs mostly from the phloem tissues of the host. Dodder seed production is prolific and seed are readily disseminated by birds feeding on the fruit, by movement in irrigation waterways and many other ways. Following initial invasion, dodder spread is very rapid as the parasitic stems twine over, between and within host plants, and may rapidly spread throughout the field. Eradication is made with non-selective herbicides or by thorough burning of host and parasite. Even small portions of the parasitic plant are capable of rapid re-establishment, growth and spread. Crop rotations can be useful to avoid dodder. Because dodder seed can occasionally be mixed with crop seed, thorough seed cleaning will reduce seed as an

Fig. 6.15. Dodder (*Cuscuta* spp.) infestation of carrot plants.

infection threat. Species of *Orobanche*, known as **broomrape** are other chlorophyll-less parasitic plants. Their slender branching leafless stems also enter and parasitize root tissues, and like dodder, produce numerous small seed.

INSECT PESTS

Umbellifer crop damage caused by insect pests results from insect piercing, sucking, chewing, boring or contamination of plant tissues. Insect-caused injuries also permit entry for disease organisms, and some insects are plant virus vectors. Although the number of different insect pests attacking umbellifer crops is extensive, relatively few are involved as a major cause of crop damage. Important groups of insect pests attacking umbellifers are lepidopterous larvae, and larvae of *Diptera* species, and species of aphids, bugs, beetles, weevils and leafhoppers.

Insect pest management in the production of vegetable umbellifers includes the use of chemical pesticides, and cultural and biological practices. The arsenal of insecticides varies with regard to specificity for the target insect and the mode of action that acts upon the insect. Some of the better known insecticides used for carrot, celery, parsnips and parsley production are: methomyl, permethain, methoxychlor, diazinon, endosulfan, carbaryl, acephate, methamidophos, cyromazine, oxamyl and esfenvalerate. Growers and applicators should be aware of the specificity of these compounds because repeated use of some insecticides does result in the development of insect populations that are resistant to the frequently used material or materials of similar chemistry and modes of action.

Carrot fly (*Psila rosae*) probably is the most serious insect pest of carrots, parsnips, celery and parsley. The insect is present in many temperate growing regions, and causes large economic losses in Europe, Canada and New Zealand. The flies lay eggs in soil near stems and the emerging larvae move further into the soil to feed on lateral roots or burrow into and mine storage roots causing plant wilting and sometimes death (Fig. 6.16). Surviving carrot and parsnip storage roots are disfigured and unmarketable. Late-developing insects remain as larvae and continue to feed on carrot roots throughout the winter. The insects overwinter in soil either as larvae or as diapausing pupae. Those that overwinter as larvae form pupae during the spring (March/April). The flies emerge soon after to lay eggs for the next generation. The possibility of several generations developing during a growing season greatly increases

Fig. 6.16. Carrot roots infested with carrot root fly *Psila rosae*. Larvae can be seen protruding from the upper surface of the lower intact root, also from the bottom half of the upper cross section portion. Larvae feeding is also evident in upper mid portion of the top root.

the potential for crop damage. Collier and Finch (1996) reported that complete development from egg to adult fly in the laboratory required 60 days at 21.5°C, and up to 254 days at 9°C.

Insecticidal control is fairly effective, but expensive and environmentally undesirable. Most early sown crops can be protected against first generation flies by applying insecticides during drilling. If harvested before the second generation appears, these crops may escape damage. However, crops requiring a long growing season are more difficult to protect unless supplemental mid-season insecticides are applied.

Accurate timing of insecticides is essential for effective control. Insecticides used are organophosphates or carbamates. Some of the insecticides used to control carrot fly include: chlorfenvinphos, carbofuran, phorate, triazophos and disulfoton. These products, often granular formulations, are applied during seed drilling, below and to the side of the seed or incorporated by bow wave application or by soil incorporation after broadcasting and before planting. Deep vertical insecticide placement improves protection and is more effective than standard bow wave application even when less total insecticide is applied. Seed coatings containing insecticides are usually not used because of possible phytotoxicity to emerging seedlings.

Sticky and water traps have been used to monitor populations in order to identify thresholds, and to improve timing of insecticide applications if and when necessary. A forecast system was proposed for scheduling the timing of insecticide sprays using the relationship between temperature and development of carrot fly. However, using a degree-day model to forecast carrot fly activity has limitations when the different insect stages have different development thresholds and the relationships between rate of insect development and temperature are non-linear. A monitoring programme for predicting peak egg-laying periods is used to schedule plantings after first generation egg-laying has ceased and which will permit plants to be well-established before second generation attack (Ellis *et al.*, 1987).

A non-chemical approach for control is to prevent carrot fly access to the plants. Covering the crop with a perforated clear plastic film for the period of first generation carrot fly activity serves as a protective barrier and is an effective method of reducing damage (Ellis and Hardman, 1987). Non-woven fine mesh polypropylene row covers are also occasionally used. In addition to excluding the flies, such row covers alter the crops micro-environment that can result in earlier and greater productivity. However, that environment is also conducive to enhanced weed growth which if not eliminated will reduce yield (Peacock, 1991).

Biochemical analysis using HPLC identified chlorogenic acid as the main phenolic acid in carrot peel (Cole, 1985). An inverse relationship was shown between the concentration of chlorogenic acid detected in carrot peel and field resistance to carrot fly damage; and the correlation held throughout the season (Ellis *et al.*, 1984; Cole, *et al.*, 1987). The location of the phenolic acid was confirmed by detection of ultraviolet fluorescence. Low fluorescence indicated a higher level of resistance. Use of fluorescence screening of carrot seedling radicles for resistance to carrot fly also correlated well with fluorescence of seedling plants.

Highly significant correlation between levels of carrot fly injury and the concentration of chlorogenic acid in tissues has permitted formulation of a predictive model of the relative predisposition to carrot fly larval damage. However, the model's accuracy is affected by environmental factors, previous attack by carrot fly, and overwintering of roots, each of which affect chlorogenic acid concentrations. For example, early season carrot fly damage increases chlorogenic acid concentrations and predisposes roots to further fly attack and damage.

Plant breeding directed towards reducing the formation of chlorogenic acid, and thereby increasing resistance to carrot flies, could alleviate this pest problem. Moderate resistance exists in cultivar 'Sytan' (low chlorogenic acid content), whereas cultivar 'Danvers 126' (high chlorogenic acid content) exhibits high susceptibility. The cultivar 'Flyaway' resulted from selection for carrot fly resistance. Additional cultivar resistance is anticipated from breeding progress. An integration of pest management and a greater level of genetic resistance in some carrot cultivars has helped to limit or avoid damage, while also reducing the quantity and frequency of insecticide applications.

Leafminers, *Liriomyza sativae, L. trifolii* and *L. huidobrensis*, respectively, **vegetable**, **serpentine** and **pea leafminers**, are present in many production areas, and attack many umbellifer vegetables. Adult insects cause striping of leaves, and larvae mine leaf mesophyll tissues which interferes with photosynthesis. Leaf mines may also provide entry for disease organisms. Primary control relies on chemical pesticides, although leafminers do exhibit rapid resistance development. Several hymenopterous parasites are useful in providing some control without themselves causing economic damage. The development of leafminer resistance for celery is being pursued through plant breeding utilizing the resistance of the wild species *Apium prostratum* in crosses with celery cultivars (Fig. 6.17b).

The larvae of the **celery fly/leafminer** *Euleia heraclei*, mines the leaf tissues of celery and also attacks parsley, parsnip and angelica. The **carrot leafminer** *Napomyza carotae*, is of some importance as a pest of parsnip, parsley, other umbellifers and even wild carrot in Europe. The larvae mine leaves and may travel through petioles into the upper portions of taproots.

Whiteflies (*Bemisia* and *Trialeurodes* spp.) generally are insects of tropical and subtropical regions, but recently have emerged as significant pests in temperate region open field umbellifer production. Important whitefly species causing damage to North American crop production include **sweet potato** and **silverleaf whiteflies**, *Bemisia tabaci* and *B. argentifolii*, respectively. Whiteflies damage celery, carrot and other umbellifers by their foliar feeding and deposition of their sticky sugary honeydew-like excretions

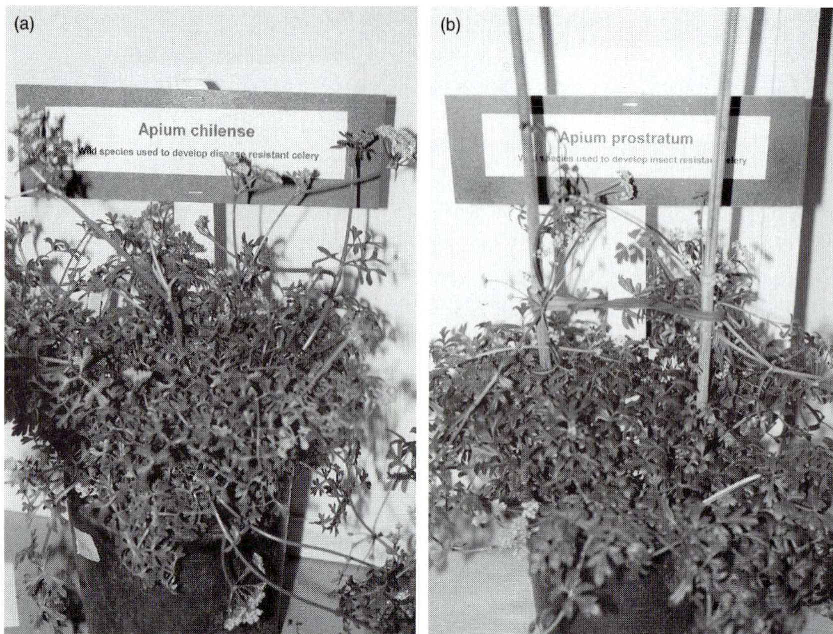

Fig. 6.17. (a) *Apium chilense*, a wild celery species conveying resistance to *Septoria*. (b) *Apium prostratum*, a wild celery species conveying resistance to leaf miner.

which soils foliage and frequently is colonized by fungal growth that produces a black sooty mould. The **greenhouse whitefly** *Trialeurodes vaporariorum*, is an important pest of umbellifer crops grown in protective structures.

Aphids are responsible for considerable economic losses of most umbellifer crops. Plants are damaged by their feeding and toxic saliva that can cause foliage deformities such as leaf twisting, cupping and gnarling. Their feeding, much like that of whiteflies, weakens the plant and also results in the deposition of a sticky sugary waste 'honeydew' which covers tissue surfaces with a varnish-like appearance that also may be a substrate for mould development. Several aphid species are virus vectors, and this aspect of aphid damage often exceeds the losses caused by aphid feeding and colonization of plants.

The **green peach aphid** *Myzus persicae*, probably is the most damaging aphid and overall major pest because it is so prevalent and is a vector for several damaging viruses. Other important aphids are: **violet aphid** *Myzus ornatus*, **pea aphid** *Acyrthosiphon pisum*, **bean aphid** *Aphis fabae*, **melon aphid** *Aphis gossypii*, **willow-carrot aphid** *Cavariella aegopodii* and **parsnip aphid** *Cavariella pastinacae*. Several others, such as: **hawthorn-parsley aphids** *Dysaphis apiifoli* and *D. foeniculus*, **hawthorn-carrot aphid** *Dysaphis crataegi* and **honeysuckle aphids** *Rhopalosiphum conii*, *Hyadaphis foeniculi* and *H. coriandi*, also cause crop production losses. Honeysuckle aphids attack coriander, anise, dill, fennel and parsley in addition to carrot and celery. The **potato aphid** *Macrosteles fascifrons*, is an occasional pest of carrot, parsnip, anise and angelica. Pesticides are generally used for control, but because of the mobility of aphids and rapid immigration of large populations, control is not always adequate. Plant variation in susceptibility to attack and aphid colonization exists among some umbellifer species and cultivars. Removal or reduction of refuges of virus-hosting weeds helps to reduce aphid disease transmission.

Tarnish plant bugs are species of *Lygus*. Those most often inflicting crop damage are: *Lygus lineolaris, L. pratensis, L. rugulipennis, L. hesperus* and *L. elisus*. Because adults can overwinter, the onset of early favourable temperatures can rapidly promote a large increase in pest populations. Their feeding and toxic saliva cause brown 'sting-like' lesions, and their piercing and feeding of developing seeds reduces seed quality and yield. It is estimated that one insect per carrot umbel can reduce yield by 20% and seed germination by 25%. Control with insecticides is effective, but is complicated during seed production by the need to eliminate the pest without harming insect pollinators.

Leafhopper species are serious pests of a number of umbellifer crops in nearly all areas where these crops are grown. Crop damage is caused by the piercing of leaf and stem tissues which kills cells and incurs disease transmission. Perhaps the most important pest is the **aster** or **six-spotted leafhopper** *Macrosteles fascifrons*, which is the primary vector of aster yellows. Leafhoppers, *M. divisus* and *Aphrodes bicinctus*, also are vectors of this disease. The **beet leafhopper** *Circulifer tenellus* is a vector of the beet leafhopper-transmitted virescent agent (BLTVA) that infects carrots. Leafhoppers are long-distance migrators and repeated infestations can occur in production fields. Monitoring in order to predict population migration and their level of aster yellow infectivity are made to improve the effectiveness of insecticide treatments.

Carrot psyllid (*Trioza apicalis*) is a frequent northern Europe pest of carrot. Like the aphid, they are phloem feeders. When feeding, psyllids inject a toxin along with their saliva that reduces plant and especially root growth, and also causes leaf deformity. Psyllid feeding also is known to reduce carrot root carotene content. Additionally, 'honeydew' deposition contaminates the foliage.

Carrot weevils, *Listronotus oregonensis* and *L. latiusculas*, occur in the north-eastern USA where they periodically inflict severe losses to crops of carrot, celery, dill, parsley and parsnip. Weevils are infrequently found in warm climates because temperatures above 24°C greatly reduce their activity. The adults lay eggs in petiole bases and on the tops of taproots. Adult damage is minor compared with those of the larvae which tunnel into plant crowns and roots reducing quality and sometimes killing plants. Late plantings can minimize damage, but usually do not provide sufficient control.

Flea beetle (*Systena blanda*) feeding causes pitting or holes in the undersides of leaves. Adult feeding by this pale striped insect stunts or even kills seedlings; older plants usually survive the damage. Larvae feeding on roots also cause serious damage. Insecticides are effective, but often are not used as they may disrupt the biological control provided by beneficial insects.

Leaf-tier (*Udea rubigalis*), best known as celery leaf-tier, occasionally is present in damaging population levels. Other species identified as leaf-tiers are *Oeobia rubigalis* and *Phlyctaenia rubigalis*. These pests feed on tender young celery leaves, namely the interior heart leaves. During feeding the insect covers the feeding area with webbing and its excrement, thus adding insult to injury.

Thrips cause crop injury with their tissue-rasping mouth parts. Their feeding causes scarring, and sometimes bronzing of leaves and petioles. The injury provides entry points for secondary disease organisms. Thrips also are vectors of Spotted Wilt Virus. Several important species are: **flower thrips** *Frankliniella tritici*, **western flower thrips** *F. occidentalis* and **onion thrips** *Thrips tabaci*. Pesticide applications are the most commonly used control practice. Additional insects known to attack umbellifer crops are listed in Table 6.2.

LEPIDOPTERA PESTS

Worldwide, several Lepidoptera insects categorized as cutworms are serious and chronic pests. In particular, the **common cutworm** or **turnip moth** *Agrotis segetum*, **black cutworm** *A. ipsilon*, **dart moth** *A. exclamationis*, **variegated cutworm** *Peridroma saucia* (Fig. 6.18) and **granulate cutworm** *Feltia subterranea* are responsible for considerable plant damage. Cutworm feeding upon all plant parts causes major damage to mature plants, and

Table 6.2. Additional insect pests reported to attack umbellifer crops.

Blister beetle	*Epicauta* spp.
Carrot beetle	*Bothynus gibbosus*
Corn wireworm	*Melanotus communis*
Darkling ground beetle	Tenebrionidae family
Field crickets	*Gryllus* spp.
Garden fleahopper	*Halticus bractatus*
Grasshoppers	*Melanoplus differentialis* Acrididae family
Potato wireworm	*Conoderus falli*
Springtails	Collembola family
Stink bug	Pentatomidae family
Symphylans	*Scutigerella immaculata*
Wireworms	*Limonius, Melanotus* and *Conoderus* spp.

Because insect common names often vary and a specific insect may be known by several different names, the Latin binomial provides the best identification.

Fig. 6.18. Variegated cutworm *Peridroma saucia*. Source: Nick Toscano, University of California, Riverside, USA.

seedlings can be killed when larvae chew through stems at the soil line, or burrow into roots. Control is obtained with several *Bacillus thuringiensis* products, chemical pesticides and beneficial predator or parasitic insects, spiders, fungal and/or insect viruses. Granulosis viruses applied as an aqueous suspension have provided effective control against some cutworms. The removal or early destruction of infested crops can reduce subsequent insect pest populations.

Another widely distributed lepidopteran pest of umbellifers is the **beet armyworm** *Spodoptera exigua* (Fig. 6.19). At all larval stages, the insect is a general feeder attacking foliage, stems and roots. Parasites of beet armyworm do not present sufficient control and chemical pesticides presently are the primary control materials. Other insects sometimes identified as armyworms are *Spodoptera frugiperda* and *Pseudaletia unipuncta*. Larvae of the **salt marsh caterpillar** *Estigmene acrea* also are indiscriminate feeders attacking young and mature umbellifer plants. Their control relies on insecticides. Another very widely distributed pest is the **cabbage looper** *Trichoplusia ni*. Damage is caused by larval feeding during all stages of plant growth, and insect frass deposition produces an undesirable cosmetic appearance. Control is largely dependent on pesticides, although pesticide-resistance development by this pest can develop rapidly. Predators are few and presently do not provide reliable control. The **parsnip moth/web worm** *Depressaria heracliana* occurs in Canada, Europe, Africa, and northern Asia where seed rather than plant damage is the more important injury caused by the larvae. Larvae produce

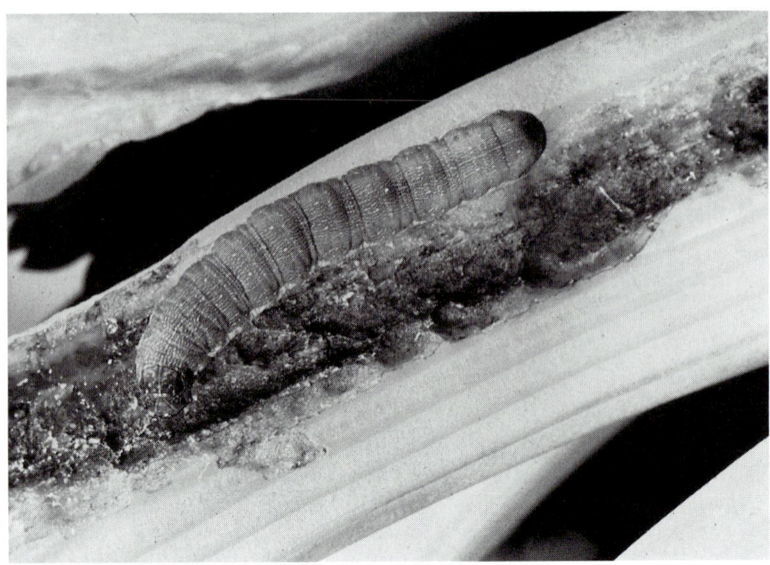

Fig. 6.19. Beet armyworm *Spodoptera exigua*. Source: Nick Toscano, University of California, Riverside, USA.

webs that cover parsnip and celery umbels and consume the umbel tissues and immature seeds. The insects also mine through stems and feed on foliage. Additional lepidopterous insects reported to attack umbellifer crops are shown in Table 6.3.

Table 6.3. Additional lepidopterous insects reported to attack umbellifer crops.

Alfalfa looper	*Autographa californica*
Celery looper	*Autograph falcifera*
Common lancewing moth	*Epermenia chaerophyllella*
Corn earworm	*Helicoverpa zea*
Diamondback moth	*Plutella xylostella*
European corn borer	*Ostrinia nubilalis*
Garden dart moth	*Euxoa nigricans*
Garden swift moth	*Hepialus lupulinus*
Ghost moth	*Hepialus humuli*
Parsley caterpillar	*Papilo zelicaon*
Soybean looper	*Pseudoplusia includens*
Whiteline dart moth	*Euxoa tritici*
Yellow underwing moth	*Noctua pronuba*

INVERTEBRATE PESTS

Mites *Tetranychus telarius* (syn. *T. urticae*) and *T. cinnabarinus*, known as spider mites, are the most important species attacking celery, parsnip, parsley and several other umbellifers. **Crown mite** *Tyrophagus dimidiatus* also is a damaging mite pest. The celery rust mite *Aculus eurynotus* is a specific pest of celery. Mite feeding causes foliage of infested crops to be light-coloured and high populations can cause tissue desiccation and death. Mites are web producers and their webs reduce market appearance. Pesticides are the primary control method, but crop sanitation, avoiding dry and dusty field environments, and crop rotations can reduce the frequency and severity of infestations.

Snails and slugs, namely the **brown garden snail** *Helix aspera*, and **slugs** such as **grey garden** *Agriolimax reticulatus* and **black European** *Arion ater*, damage crops by their feeding and because their presence on or in the commodity is undesirable. *Milax gagates* and *Deroceras reticulatum* occasionally cause economic crop losses.

BENEFICIAL PARASITES AND PREDATORS OF INSECTS AND MITES

Several beneficial insects directly attack and consume the eggs or larvae of certain insect pests and also nymphs and aphid adults. Wasp species of *Encarsia*, *Eretmocerus* and *Lysiphlebus* parasitize eggs and young larvae of umbel-damaging lepidopterous insects. Fortunately, more of these hymenopterous parasites continue to be discovered. The green and brown lacewing, ladybird, collops beetles, and big-eyed, pirate, assassin and damsel bugs attack many insect pests of umbellifer crops. Larvae of the syrphid fly specifically attack aphids. Several spider species are used as predators because they indiscriminately feed on many insect pests.

Beneficial insects are useful in crop production because they destroy pests. However, they are a potential nuisance for marketing because most consumers are not tolerant of any insects, insect parts or wastes on or in crop products. Nevertheless, the reliance on beneficial insects for biological control practice is increasing and commercial rearing of high populations and introduction of predator and parasitic insects into production fields to combat insect pests is increasingly becoming a common procedure.

VERTEBRATE PESTS

Several bird species damage crops by their consumption of planted seeds, newly emerged seedlings, and occasionally foliage and other portions of developing plants. Bird droppings also occasionally spoil crop market appearance.

Fig. 7.1. (a) Hand harvested carrots prepared for bunching, (b) master bundles consisting of 10–12 individual bunches, (c) transport of bunched carrots to packing shed for washing, cooling and packing. The load appears jumbled, but is arranged so that unloading is easily accomplished without damage.

Fig. 7.2. Mechanical harvesting of carrots. (a) The machine is capable of lifting two beds (four rows) simultaneously at a rapid rate, removing the tops, and conveying the roots into large trailers. (b) The roots are undercut and lifted from the soil by belts that grasp the tops.

producers achieve between 40 and 50 t ha^{-1}. Harvesting is often delayed until first frost or even later in order that the plants are exposed to cold conditions. Cold exposure in the field results in the conversion of some of the storage root carbohydrates into sugars, which is noted as enhanced root flavour and sweetness. When low outdoor temperatures are insufficient, short-term (10–14 day) holding at cold temperature, preferably at 0°C, can increase root sweetness.

The roots are harvested much like those of carrots, being undercut and then lifted and deposited on to the soil. Careful handling is required to avoid damaging the thin surface tissues of the root. Hand trimming of the foliage is somewhat difficult because the stem base is sunken below the shoulders of the parsnip root, and crowning practices are often utilized. When marketed as bunched parsnips, the plants usually are harvested at an earlier stage of growth, and several roots of similar size are grouped and the foliage tied together. Lateral roots, if present, are broken off or trimmed. The strongly tapered narrow lowest portion of the root is most likely to exhibit rapid desiccation and is often trimmed away.

Celeriac harvesting

Sometimes known as celery root, turnip-root celery or knob celery, celeriac is generally grouped as a root crop, although in botanical terms the product is an enlarged stem–root axis. The crop has a relatively long growth period, ranging upwards from 4 to 5 months after planting. Yields of 35–40 t ha^{-1} are obtained by good producers. Harvest delays can result in the development of tissue softening 'pithiness' of the enlarged organ, and occasionally the formation of a hollow area in the upper portion of the root below the stem.

Celeriac machine harvesting involves undercutting and ploughing out or lifting the plant. In smallholdings or home production, the plants are simply hand dug. The tops may be mowed before lifting or used to lift the crop from the soil. Preparing celeriac for market presentation involves considerable trimming of foliage and of the numerous inter-twined lateral roots, which is accomplished by hand labour or with machinery. Mechanical trimming is performed in a rotating drum that tumbles the celeriac against abrasive surfaces that grind away the many small lateral roots, portions of the attached petiole bases and some basal leaf scars to produce a relatively smooth globe-like product. Washing before trimming only partially removes the ever-present adhering soil, and a thorough washing is usually required after trimming, or during the tumbling action of mechanical trimming.

Arracacha harvesting

The growth period for the typical market size development of arracacha roots is 10 months or more after planting, although the storage roots can be harvested as early as 6–8 months. Yields range from 5 to 15 t ha^{-1}. Harvest usually begins just before plants flower and when leaves first begin to senesce and yellow, although the status of root development can be examined directly by careful removal of surrounding soil. Early harvest compromises yield, but harvests delayed beyond the flowering period result in fibrous and poorly flavoured roots. The plant is injured by exposure to low temperatures, and is intolerant of frost.

Harvest involves uprooting the entire plant, either by hand or mechanically. The lateral roots, usually four to six, and sometimes as many as ten, are carefully pulled away or cut from the central root. It is important to avoid damage of the easily bruised root surfaces, and to minimize desiccation. The celery-like stem and succulent petioles for vegetable use are also hand harvested, and generally used fresh within a few days.

HARVEST OF OTHER UMBELLIFER ROOT CROPS

Significant production of **Hamburg parsley** occurs in Central Europe, although the crop is not an extensively grown crop in other locations. When production volume justifies, the roots can be machine harvested much like carrots, and with similar equipment. However, it is common that the plants are undercut and hand lifted from the soil. Typically a number of plants are tied together in a bunch, or the roots are bulk handled after the leaves are topped. **Turnip-rooted chervil** is harvested and handled much like Hamburg parsley, whereas the clustered storage roots of **skirret** (Fig. 2.9) make harvest more difficult. Hand digging is common for this and other crops that are produced in small volume. The extracted roots are brushed free of attached soil, but generally are not washed before marketing.

Unlike other umbellifers, the harvested product of **asafetida** is the dried coagulated resinous extrusions from the root surfaces. This material which is produced by the roots during growth is scraped off and collected by hand. During plant growth the soil surrounding the root is moved away in order to make several deep incisions in the root. The soil is then replaced and about 30–35 days later the soil is again moved away from the roots, and the resinous exudate harvested. Additional incisions are made into the root surfaces, and the soil is replaced to repeat the procedure.

HARVEST OF LEAFY UMBELLIFER CROPS

Crops harvested for their leafy portions, succulent petioles and stems such as cilantro, celery and lovage are highly susceptible to injury and desiccation. In those few situations when the crop is produced for processing, harvest machinery is used. Even when the scale of production is large, most foliage umbellifer vegetables are hand harvested and trimmed.

(a)

(b)

(c)

(d)

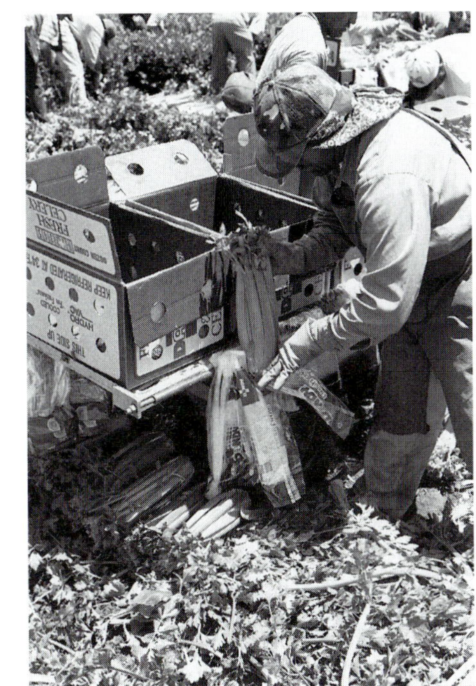

Fig. 7.3. Celery hand harvesting. *Opposite*: (a) Plants are cut and the stem bases (butts) are trimmed, (b) the upper foliage is trimmed and (c) placed on the ground for pickup by packing labourers and (d), *above*, packed in plastic film sleeves, or directly into water-resistant paperboard cartons.

Celery harvesting

Growing conditions have a strong influence on time and rate of development as does choice of cultivars. Most celery cultivars are harvested about 75–100 days after transplanting, although poor growing conditions can extend the growth period. Some cultivars normally require a longer growth period before they are suitable for harvest. Nevertheless, it is characteristic that the daily growth rate of celery plants increases very rapidly near the optimum stage of development before harvest (Fig. 5.7). Fresh weight yields of trimmed stalks can range between 60 and 90 t ha^{-1}, and in some production situations it has been observed that as much as 2–3 t of fresh weight day^{-1} ha^{-1} was produced by the celery crop. Accordingly, over-maturity is also reached rapidly, with accompanying rapid quality loss. Harvest scheduling becomes an important management decision in order to obtain both high yield and high quality, because the suitable harvest period is relatively short and not as flexible as it can be with carrots. With advancing maturity, the outer celery petioles begin to senesce, lose firmness and therefore are often removed during trimming, which when removed results in substantial marketable yield losses. The usual trimming for market-ready presentation of celery reduces stalk weight from 5 to 20%. However, when necessary to remove damaged or diseased outer petioles such trimming can range from 20% to as high as 40% of total stalk weight. Additionally, because of their succulence, celery plants are very susceptible to injury, and damage adds to any necessary additional trimming.

Although machines are occasionally used to harvest celery for processing, most of the celery produced is harvested with hand labour. The roots are cut below the soil surface and below the base of the stem. The severed stalk is picked up and its overall length is reduced by trimming away much of the uppermost leaf blades with a knife, and damaged or otherwise unmarketable petioles are removed (Fig. 7.3a,b). Often a small portion of the upper root and lower stem is retained to help reduce wilting during postharvest handling and storage. This also allows for further trimming of the stem base to present a fresh appearance. The harvested stalks are collected and placed in boxes, bulk containers or trailers for transport to packing facilities or packed in the field (Fig. 7.3c,d).

Many approaches have been investigated in order to mechanize the harvest of fresh market celery. Partial mechanization is provided by machine mowing of the uppermost leaf blades to reduce that aspect of hand trimming, and by undercutting the plants to ease removal from the soil. At one time large machines called 'mule trains' were used in the field to assist labourers with the harvest of celery and as a mobile substitute for a stationary packing shed. Some versions of these machines were self-propelled and were used to mow the leaf tops, lift and convey the stalks to the workers who washed, trimmed and packed the celery. The containers were also constructed on the machine. These machines carried as many as 50 workers. The use of these machines was discontinued because their large size made them cumbersome and heavy, and their weight compacted the soil, especially when they carried water for washing. Their size made them difficult to be moved from field to field or to distant production areas, and their use was not efficient unless fields were large. Furthermore, breakdowns or any other stoppages idled many workers until the operation could resume.

The use of smaller harvest machines to cut and lift celery had the serious drawback of delivering a jumble of disoriented and damaged stalks that only complicated trimming and packing efforts. When compared with hand harvest and handling, these machined did not save labour or expense.

Considerable amounts of celery for canning or dehydration processing are harvested entirely by machine. The harvesters, essentially simple mowing machines, are adjusted to cut the plant slightly above the ground and convey the celery into bulk bins or trailers for transport to the processing facility. Although many individual petioles are detached from the stem, and jumbled in the harvest containers, this is less of a problem for processing than for fresh market use.

Smallage harvesting

Smallage plants resemble stalk celery, except that their growth period is shorter, about 75–90 days after planting. The plants are smaller, and petioles less thickened, although high-density plantings and favourable growing conditions can produce yields comparable to those of stalk celery. The crop is harvested by hand labourers much like stalk celery. The root is cut just below the stem and except for removal of unusable petioles, the plant is usually left intact. If the growth period is extended the plants become larger, but edible foliage quality suffers. Being smaller than celery, several plants commonly are bundled together to facilitate postharvest handling (Fig. 7.4a,b).

In another harvest practice the foliage is cut above the growing point to permit leaf re-growth and an additional harvest. Small-scale growers also perform multiple harvests by removal of several of the largest outer petioles and allow remaining petioles to develop further for later harvesting. Because of marketing and handling practices common for smallage, mechanization of harvest offers little labour reduction. Additionally fresh market production of smallage on smallholdings cannot justify the investment into harvest equipment. However, as grown for processing, the crop is well-suited to machine harvesting.

Parsley harvesting

Optimum harvest yields of leaf parsley generally are achieved about 90 days after planting. Earlier harvests of plants at less than full development are commonly performed by fresh market producers. Much of the fresh market crop is presented as hand-tied bunches. At harvest the plants are undercut slightly below the stem base which keeps the foliage attached. Several plants of similar size that meet market size or weight specifications are bunched (bouquet style) and the petioles tied together (Fig. 7.5a). When plants are larger, the foliage is also hand cut, and the leaves are grouped together and tied as bundles. In this procedure, the leaves are often cut slightly above the growing point to allow for re-growth and additional harvests.

A considerable volume of parsley is produced for dehydration processing. Harvest machines are commonly used for this, and the procedure is fully mechanized (Fig. 7.5b). The machine's cutting mechanism is adjusted to cut the foliage above the stem apex to permit re-growth, and also to reduce the ratio of petiole to leaf blade tissue, which is an important parameter for dehydration processors. Parsley's growth habit permits multiple harvests, and in locations where temperatures are not conducive to bolting, crop growth is continued for more than one growing season.

Fig. 7.4. (a) Smallage bunched as prepared for fresh market. (b) Handling of smallage in a Chinese market (note use of ice in container to retain postharvest quality).

(a)

(b)

Fig. 7.5. (a) Parsley hand harvesting and market preparation as bunches. (b) Mechanized harvesting of parsley for dehydration processing.

Harvest of other leafy umbellifer crops

Cilantro foliage is harvested and handled much like parsley, with either several intact plants or cut foliage bundled and tied together. Plants may also be cut so as to allow re-growth and repeated harvesting. Plants are harvested about 40–50 days after sowing, although some plants grow for 70 days before harvests are made. Fresh weight yields of 10 t ha^{-1} are obtained from three or four harvests.

The harvest of the bulb-like leaf petiole bases of **Florence fennel** generally occurs about 80-plus days after planting. Sometimes plants are grown at high densities to produce 'baby' fennel which is harvested at an earlier growth stage, usually about 60 days. Mechanical undercutting of the plants is occasionally practised, but usually the taproot is hand cut just below the stem base so that the plants are easily lifted from the soil (Fig. 7.6a). The fine leaf blades and much of the upper portions of the petioles are trimmed away before field packing (Fig. 7.6b,c). Petiole trim length varies considerably with market requirements, but ranges between 15 and 30 cm. The upper petiole portions and fine leaves are occasionally utilized as a potherb, and these can be harvested at almost any time for fresh use, although sparingly so as not to significantly affect development of the enlarging petiole bases.

Chervil is usually harvested about 50–60 days after planting. Harvest and handling is similar to that for parsley, and repeat harvests are possible. The seldom-cultivated **cow parsley** also resembles chervil, although the leaves tend to be larger. **Lovage** plants also have many similarities to celery, although rather than the entire plant being harvested, only the outermost leaves are detached, and bunched together for marketing. The succulent stems, which often are blanched, are also harvested for fresh use. Roots are dug in autumn or early spring, often in the second or third season of growth when significant enlargement has been achieved.

Japanese hornwort plants are generally harvested as intact plants with attached petioles and leaflets, usually when petiole length is between 15 and 20 cm. The petioles tend to be longer (30 cm or more) when the crop is blanched. The plant is detached by cutting through the root just below the stem. The petioles resemble those of smallage (Fig. 7.7). In an alternative harvest method, the outermost petioles can be pulled off and bundled together for marketing. The remaining less-developed petioles continue to grow and enlarge, and thereby the harvest period is extended and final yields increased. The almost prostrate foliage of **Asiatic pennywort** plants essentially mandates hand harvesting. Harvests can be repeated every 50–60 days following re-growth. Aquatic conditions limit harvest mechanization of **water dropwort**. As with most other leafy crops, the leaves and young stems are hand cut, and bundled in preparation for marketing which typically occurs in the fall and winter. **Garden angelica** foliage is cut before flowering occurs. Stems and petiole bases are harvested when young and tender. The roots, if of adequate size, are dug in the fall, or allowed to enlarge during a second or third year.

Fig. 7.6. (a) Fennel hand harvesting, (b) packing operation and (c) packaged product.

Fig. 7.7. Japanese hornwort plant.

HARVEST OF UMBELLIFER SEED CROPS

Carrot and celery seed harvesting are discussed in Chapter 3. The scale of carrot and celery seed production usually justifies machine harvesting, but production volume usually determines whether other crops grown for their seeds are hand or machine harvested. The harvest procedure for most other umbellifer seed crops often begins early in the day since the presence of dew is considered advantageous in order to minimize seed shattering. In some situations combine machines harvest and separate seed in one operation, although seed yields tend to be lower with single operation combine harvesting. Alternatively, seed plants are cut or pulled from the soil. The loosened plants or inflorescences are windrowed in the field or piled on to ground covers for further maturation and drying. Protection against rain or contact with moisture is important in order to avoid loss of seed quality. Ground covers protect plants from soil moisture and also capture seed that may have scattered.

Thrashing may occur in the field when the windrows are picked up or the plant material is carried to stationary thrashing units. When seed volume is small and machine thrashing is not feasible, hand thrashing is commonly performed. With few differences, the seeds of most umbellifers are similarly harvested and handled. In order to maximize seed yield the mature primary umbels may be harvested first, followed by a second harvest for later-developing umbels. The usual indication of appropriate maturity for most seed crops is a change in fruit or seed colour from green to brown.

Coriander, although a fairly substantial condiment crop is usually hand harvested because machine harvesting results in damage to the fruit containing the mericarps (seeds). When damaged, the fruit quality is compromised. The harvest is usually scheduled for about 100 days after planting as the seed approach maturity, but before significant shattering occurs. Coriander crop yields range between 1 and 2 t ha^{-1}, occasionally approaching 3 t ha^{-1}.

POSTHARVEST HANDLING AND PACKING

Important postharvest procedures begin with the timely harvest at an optimum period of crop development or maturity, and careful handling throughout. Minimal and careful handling during and following harvest reduces physical injury of the harvested product. Rapid postharvest cooling of root and foliage crops, and appropriate relative humidity maintenance minimize quality loss (Fig. 7.8). Harvested products are cleaned, trimmed, washed and/or graded as needed, and packaged into containers that facilitate cooling and handling, and protect the product from damage.

Because they are prone to desiccation, it is especially important that leafy and stem umbellifers are rapidly cooled following harvest. Lowered temperatures reduce product respiration and desiccation and, thereby loss of

Fig. 7.8. Hydro-vac vacuum cooler. During the cooling procedure, water is introduced to limit moisture loss from the product and the appearance of desiccation.

weight. Umbellifer root crops are somewhat less susceptible to rapid desiccation, but low temperatures are also important to minimize tissue respiration to avoid weight loss and wilting.

Desiccation is not a concern for harvested seeds but inappropriate postharvest handling, temperature and relative humidity conditions will affect viability and quality. The postharvest handling of umbellifer seed crops often involves a drying period to facilitate handling and lengthen shelf-life. This is followed by cleaning to separate the seed from plant trash, mainly umbel parts, and the removal of other contaminants. During cleaning the mericarps become separated. Associated operations remove spines, size grade and dry seed to the moisture content appropriate for each species. Seeds to be used for propagation may receive additional handling procedures to improve performance and storage characteristics. Seeds produced for condiment purposes after cleaning and drying are maintained in dry and cool conditions, preferably with low and controlled relative humidity. Such seeds are often held in sealed containers to minimize the loss of aromatic and other flavour compounds.

Carrot postharvest handling

Following harvest, fresh market carrot roots, with or without attached foliage are transported from the field in bulk containers or wagons to packing sheds. There they are unloaded, often into a water tank to reduce impact damage and remove attached soil. Bundles of carrots with attached foliage are then washed with clean water. Having been graded before bundling they can be packaged directly into waxed or film-coated water-resistant fibreboard containers. Because of low availability and high cost the use of wood containers has greatly declined. The packed carrot bunches are often accompanied with shaved ice for additional cooling and to limit desiccation. Usually the product is maintained in a cool train throughout handling. However, even with a temperature of 0°C and 100% relative humidity, acceptable market appearance is limited to about 2 weeks because of eventual foliage deterioration.

Foliage-removed (topped) carrots are conveyed from the dump tank on to endless belts or in water flumes through a number of stages that include further washing, size grading, hydro-cooling and packing. Roots are size graded before hydro-cooling to avoid cooling product that might be discarded.

Cold water, preferably close to 1°C, quickly reduces root field temperatures, although the time required varies with initial temperature, product volume and the capacity to provide cold water. The small surface area of the root relative to its bulk is an impediment for rapid cooling. Nevertheless, for carrots, parsnips and other root umbellifers, hydro-cooling is more effective than other cooling methods. Hydro-cooling also provides the benefit of some rehydration of slightly wilted roots. If roots are not sufficiently pre-cooled, respiratory heating and moisture loss will increase and result in reduced shelf-life. After pre-cooling the roots are conveyed to packing stations for additional inspection of market suitability. Roots are packaged according to market requirements in bulk or pre-pack type containers (Fig. 7.9). Root size frequently is the determinant as to what kind of packaging is performed. Large roots are less suitable for pre-pack marketing, and are usually packed into larger containers. Commonly these are net or plastic film bags containing about 20–25 kg.

For pre-pack packaging, workers using scales with tilt hoppers place the number of roots that meet a specified weight into polyethylene ('cello') bags (Fig. 7.10). In the USA, individual cello packages commonly contain between five and nine roots that weigh slightly more than 454 g (about 1 lb). Multiples of other weights are also packaged. A number of such packages are consolidated into a master container to facilitate handling. These are maintained in cold holdings during transport or short-term storage until marketed.

At some facilities the packing operation is automated; the bags are formed by the machine from a continuous roll of plastic film, that simultaneously weighs, deposits and seals the roots in the bag. The packages may have small perforations for ventilation to avoid the development of unfavourable odours and flavour.

Cello packaging has proven to be an extremely successful marketing practice because the carrots are protected from dehydration, and with temperature maintenance at 0°C and 98–100% relative humidity, quality can be maintained successfully for 1 or 2 months or more. Cello-bag presentation has resulted in wide consumer acceptance and confidence in obtaining a clean and easy-to-use product.

Carrots harvested for processing are handled in pallet bins and bulk trailers. They are washed, size graded and prepared according to their designated processing purposes. Forms of preparation include: canning, freezing, dehydration, pickling and juicing, and in forms that may be slices, diced portions, puréed, intact small roots or portions of roots. For some carrot juicing operations, purée is frozen in large blocks and can be stored for long periods in this form.

A relatively recent development in carrot marketing is that of minimal or lightly processed roots. For this product, high plant densities are used to produce slender roots which, after conventional harvesting, are cut into sections about 5 cm in length, accurately size graded for diameter, and the root sections are abrasively lightly peeled. The root ends become rounded and surfaces made smooth with specialized ('polishing') equipment (Fig. 7.11). Following these procedures the root section have the appearance of what was commonly referred to as 'baby' or finger carrots, which actually are small whole carrots. These root segments are packaged in film bags much like cello carrots with cold chain handling throughout. With good low temperature management,

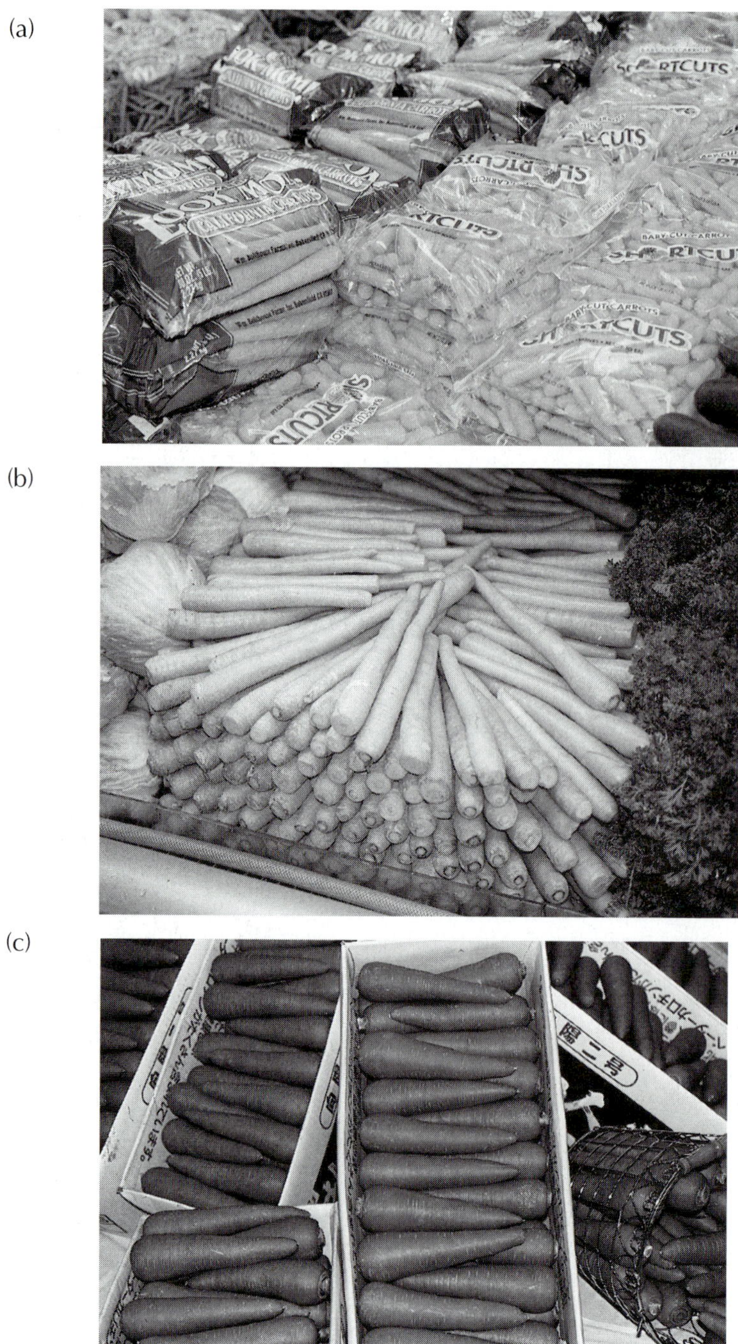

Fig. 7.9. (a) Examples of fresh market presentation of carrots in pre-packed cello bags (b) individual trimmed roots in bulk display and (c) as a carefully selected presentation for up-scale markets.

Fig. 7.10. Carrot hand pre-packing into film (cello) bags. This procedure has been replaced by automated equipment.

shelf-life can be maintained for several weeks. The excellent reception by consumers for this value-added product suggests a likely expansion of its production.

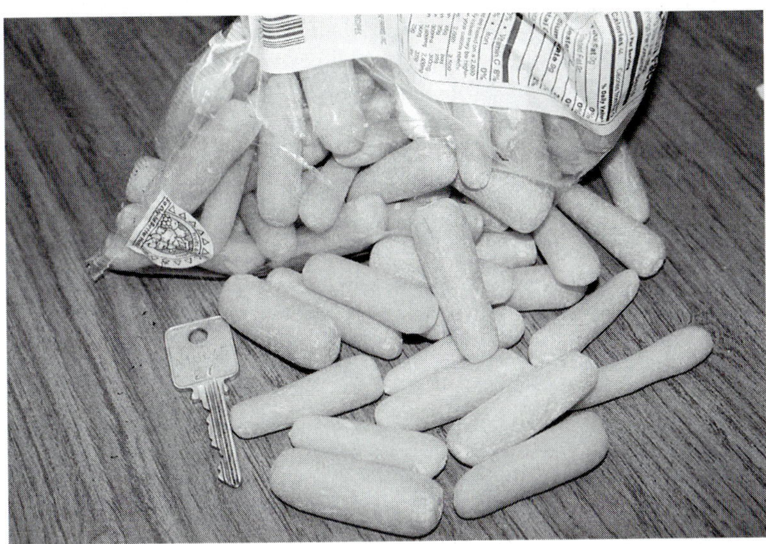

Fig. 7.11. Carrots as a highly processed value added product, frequently known as baby, short cut, or cut and peel carrots.

Postharvest handling of other umbellifer root crops

Careful postharvest handling is critical for parsnip because the thin surface layer of roots rapidly discolours and becomes brown when injured. It is also important to avoid desiccation which first becomes evident at the tip of the taproot. To limit desiccation, relative humidity must be high, preferably greater than 98%. Roots are sometimes waxed to limit wilting. However, waxing does not markedly reduce moisture loss and may increase decay. Parsnips often are packaged in cello bags much like carrots, and also bulk packed in bags or fibreboard boxes. When properly handled and stored, good root quality can be maintained for 4–6 months. Ideal market presentation provides roots that are firm, smooth, uniform in size and shape, properly trimmed and free of decay, injury, growth cracks and secondary roots.

Low temperature and high relative humidity are prerequisites for retaining postharvest quality of all root umbellifers. Like parsnip, waxing of trimmed celeriac offers little benefit to quality maintenance and increases the potential for decay. Hamburg parsley, turnip-rooted chervil and skirret have similar postharvest requirements as the other umbellifer root crops. It is important for the harvested roots to be quickly cooled and ideally maintained at 1 to −1°C with more than 95% relative humidity. Hydro-cooling is very effective for initial cooling. Thereafter, crushed ice and refrigerated cooling is useful to maintain low product temperature. The shelf-life of these crops is extended if the foliage is removed. Arracacha roots are very susceptible to physical injury and desiccation, and even with the best of postharvest conditions shelf-life is relatively short. Low temperatures of about 5°C arrest deterioration. Roots of angelica and lovage after digging are washed, air-dried and stored at low temperatures in tightly closed containers to minimize loss of aromatic compounds.

POSTHARVEST HANDLING AND PACKING OF LEAFY AND STEM UMBELLIFER CROPS

Celery postharvest handling

After harvest, if not field packed, the celery crop is transported to packing sheds. At the packing shed, the celery is unloaded for washing, additional trimming and packing into wooden boxes or moisture-proof fibreboard containers. Typically ice was added to the packages during distribution or cold room holding. Two important purposes of the packing sheds were washing and cooling using hydro-cooling, ice or vacuum cooling. Washing was necessary when blanched celery was produced in order to remove soil from the inner portions of the stalks. Thus, packing sheds met this requirement. Prior to the wider use of refrigeration, ice was the primary method of cooling and the packing shed facilities were best adapted to providing ice and its application. Field packing of green celery excludes the washing and icing that was performed in packing sheds. Package icing has been effectively replaced by hydro-cooling, vacuum and mechanical refrigeration.

In-field packing has become a common procedure for celery, fennel and several other crops, and is almost exclusively done in the USA. The change from shed to in-field packaging was a response to the higher costs of shed operations compared to field packing. Small movable wheeled platforms are used as field packing stations (Fig. 7.6b). In some operations, conveyors are used to assist handling and delivery of cut stalks to packers.

Preparation of cut stalks for packing involves trimming the stalks to a length that the packing container can accommodate; usually between 35 and 45 cm. Additional trimming strips away unmarketable and damaged outer petioles, and sucker growth. The stalks are laid on the ground or in piles for other workers to pack according to size (stalk diameter) into wood, wire-bound or water-resistant corrugated fibreboard boxes (Figs 7.3d and 7.6c). Size grades are subjectively determined, and experienced workers produce a relatively uniform product. Individual stalks of uniform size are placed parallel to each other in alternating layers in the containers. When packed, celery containers are of similar weight but can differ in size and number of stalks contained. Thus, a package of 24 large stalks would have a weight equivalent to a similar package that may contain 48 small stalks. Moisture-resistant fibreboard boxes are used in order that packed celery containers sustain contact with water. When packed, the boxes weigh between 25 and 30 kg, whereas when previously shed packed into wooden crates the containers often weighed more than 40 kg, especially if ice was added. An industry trend is towards packages of lower weight.

For some markets, individual stalks are placed into sleeve-like polyethylene bags to retard moisture loss, and improve market presentation (Fig. 7.3d). 'Hearts', a term used for under-sized stalks or those made small because of excessive trimming, are packaged as two or three stalks in a film bag. These stalks, consisting of very tender petioles, occasionally command a premium price in the market.

Postharvest cooling

Celery is usually cooled soon after packaging. Cooling effectiveness is influenced by the packing container as well as the method of cooling. The initial field temperature of the product influences how long cooling will take. Cooling rates and uniformity vary with cooling facilities and practices.

Cold room cooling is the least expensive method, but requires more time. However, it is useful for holding product that is already well cooled. Forced air cooling is more rapid, and requires continued maintenance of high humidity during cooling. Hydro-cooling with clean water, preferably with 1°C water is very effective, and although the immersion time required varies, the process is fairly rapid. Hydro-cooling also can refresh slightly wilted stalks. Hydro-vacuum cooling is the most efficient and rapid, but also is the most expensive procedure. Hydro-vacuum cooling minimizes the weight and moisture loss common with conventional vacuum cooling, because moisture for the evaporative cooling process is introduced during the cooling cycle. Thus, less water is removed from the product being cooled, whereas in the conventional vacuum procedure the source of the evaporating water is the product. Moisture loss of 5% will result in noticeable wilting of celery stalks. Being less effective, and adding considerable weight and inconvenience during transit, icing is infrequently used.

Postharvest handling of other leafy umbellifer crops

Smallage should be handled much like celery, although in many parts of the world where it is an important vegetable, it is not given appropriate postharvest attention. For parsley, good postharvest quality can be maintained for 8–10 weeks at 0°C. High relatively humidity is essential to prevent desiccation. Packaging in perforated polyethylene bags and the use of top ice is beneficial, although direct contact of ice with the foliage should be minimized. Similar conditions are appropriate for leafy commodities like cilantro, culantro, chervil, Japanese hornwort and others whose foliage is intended or prepared for fresh market. Florence fennel petiole bases are less likely to rapidly desiccate and are more closely associated with handling practices for celery. However, dill weed, and other umbellifers having finely divided delicate foliage, are very susceptible to desiccation.

POSTHARVEST HANDLING OF SEED, CONDIMENT SEED AND PLANTS

Cleanliness, dry and cool conditions, and well-sealed containers are most appropriate for postharvest seed handling whether for propagation or condiment use. Condiment use of dry leaves and roots of lovage, angelica and other umbellifers are best handled at cool temperatures, avoiding direct contact with moisture.

UMBELLIFER CROP STORAGE

Crop storage is performed to retain edible and marketable quality when marketing or immediate use are not suitable. Storage is used to: accommodate surplus production in order to manage market distribution; to extend marketing periods; and/or to await favourable marketing opportunities. Low storage temperatures reduce respiratory heating, and retard senescence and quality deterioration due to loss of dry matter and moisture. Accordingly, clean storage facilities that provide well-controlled management of temperatures, relative humidity and ventilation best serve most umbellifer vegetables.

Carrot storage

Since the 1950s, the availability of year-round production has greatly reduced the magnitude of long-term postharvest storage of carrots in the USA. In many other carrot production regions of the world harvest periods are relatively short, and some may be limited to only 1 or 2 months. Therefore, storage is often necessary to achieve a favourable distribution period, year-round if possible, for marketing of the crops produced.

Low-temperature pre-conditioning before long-term storage improves shelf-life as roots show significantly less weight loss and retain a brighter orange colour than unconditioned roots. Pre-conditioning likely enhances deposition of suberin on the surface of the periderm and lignification of the sub-surface cells, which are possible mechanisms for reducing root weight loss and discoloration.

Primary storage practices are indoor common or controlled temperature storage, and to some extent *in situ* or overwintering storage. Common storage utilizes cold ambient air and appropriate ventilation for cooling, whereas mechanical refrigeration produces the low temperatures that provide cooling.

When possible, harvests should be made when conditions minimize the amount of soil adherence on the roots, thus muddy field conditions during harvest should be avoided. Pre-storage washing may be desirable if roots are harvested under wet conditions as many potential decay-causing organisms can be washed away. Water used for washing may include chlorine to retard development of storage rot. Clean roots enhance air circulation that removes respiratory heat and reduces possible moisture condensation.

Generally it was preferable not to wash carrots before entering long-term storage, and in some situations the roots were not washed after storage. That practice has largely changed since most present-day markets demand washed roots. Rapid cooling as soon as possible after harvest is important for long-term storage. Late season production benefits from relatively low root field temperatures at harvest.

Preharvest conditions which reduce storage durability include stress, especially from excessive moisture or heat exposure, and root 'immaturity'. It is generally recognized that full season-grown roots store better than those that are harvested at an earlier time or growth stage. Clean, disinfected storage containers and structures, and avoidance of diseases and physical damage during all harvesting and handling procedures, will minimize potential storage losses. Some physical injury is unavoidable during mechanical harvesting, and efforts should be made to minimize such damage. It is also known that mechanically harvested roots do not store as well as hand-harvested carrots. However, present-day production economics make hand harvesting of any significant volume of storage carrots unlikely.

Stored roots are typically held in simple piles on the floor of the storage chamber or in large bins (commonly one cubic metre). When bulk piles are used they must not be too deep or large as crushing injury will occur and aeration within the piles is reduced. Air movement of 7–10 cm s^{-1} is usually adequate when storage temperatures are low. In addition to cool air circulation, ventilation removes CO_2 which at elevated levels is especially destructive to root quality. Relative humidity in the storage chamber should be greater than 95%, and even then some root moisture loss occurs. Varying somewhat with cultivars, roots exhibit wilting when moisture loss exceeds 5–8%. Plastic film-lined bins are frequently used to reduce dehydration. However, if infected roots are stored, decay is likely to be more prevalent in lined bins than in unlined. Fungicide treatments can be applied to reduce the incidence and severity of decay. It is generally not advisable to store carrots in bags that impede air flow. Ethanol accumulation in roots results in situations of anaerobic respiration. Exposure to any sources of ethylene in or near the storage atmosphere should be avoided to prevent the development of bitter flavour.

In some production locations where the winter weather is not exceptionally harsh, storage of carrot roots *in situ* has been practised for centuries (Fig. 7.12). The practice continues to be of some significance, most often in the UK and France.

For successful storage it is important to make the correct choice of cultivar, sowing date, covering material, time of covering and length of storage. Generally, the crop intended for storage is sown towards the latter part of the season to minimize the age of roots removed at the conclusion of storage in the spring. However, the crop must have attained sufficient growth to be 'mature', but not over-mature, with a high proportion of roots within the desired size range and with freedom from diseases, insects and other injuries (Runham *et al.*, 1992).

Freeze-preventive materials are applied before the first frost is anticipated, but if possible delayed for soil temperatures and thereby root respiration to decline. Materials used exclude light and provide insulation against low-temperature freezing, and minimize heating that would encourage shoot regrowth when warm spring temperatures occur.

Commonly used materials are a black polythene film applied over the top of the plants followed by layer of straw. The depth of the straw is about 15 cm which is equivalent to 30 t ha^{-1}. Chopped straw is easier to apply and to incorporate into soil after harvest than unchopped straw, but chopped straw

Fig. 7.12. Soil cover for overwintering of carrots in the podder soil region of Mont Saint-Michel, France. This practice has been replaced with the use of straw mulching.

has a lower insulation value. Symptoms of freezing tissue injury are lengthwise cracking and blistering resulting from the formation of ice crystals beneath the surface. After thawing, the surface tissues become water-soaked and darkened, and susceptible to bacterial invasion and decay. In areas with more intense cold, occasionally straw coverings up to 100 t ha^{-1} are used. Areas experiencing snowfall can derive some benefit from the insulation snow provides.

In-situ storage continues to be practised in peat soils because peat is an effective cold insulator and easily manipulated. The peat is moved on to plants to cover root shoulders and crowns up to a 15 cm depth. However, windy weather during cool temperatures can uncover plants and result in freezing. With mineral soils, it is necessary to increase the depth of cover. Regardless of the mulching used, good drainage away from the stored roots is important.

Some producers prefer *in-situ* storage rather than temperature-managed indoor storage because they believe that carrot post-storage root surface quality, and thus marketability, is better; other producers disagree.

Field storage can involve as much as 14 months of field occupancy, and is not without labour costs for covering and uncovering and, cost of materials. The presence of disease in the crop can lead to significant losses in such storage because there is little opportunity to prevent spread during the storage period. Thus, whenever possible the period of field storage should be minimized. Typically, in-field storage incurs a higher proportion of root decay than indoor cold storage facilities. Therefore, revenues from sales of the stored product must be sufficient to justify the practice.

Another method, now used sparingly, is placement of harvested roots in field clamps or into ground storage pits. This least complicated and low input method is used in Central Europe and other locations, and most often by small volume producers. Overall, field and *in-situ* storage practices are decreasing because of high costs for labour and material and because importation of products from other locations are able to compete with the stored products.

Controlled atmosphere (CA) storage has not been shown to be of sufficient benefit for long-term carrot storage. Although thoroughly investigated, CA research has not adequately identified a combination of temperature, relative humidity and atmospheric composition that consistently results in lower storage losses than those found in well-managed refrigerated storage. In some CA storage conditions the quality was diminished, and the incidence and severity of decay increased (van den Berg and Lentz, 1966).

It has generally been reported that CA storage at about 1°C with low O_2 and CO_2 (2–6% O_2 and 3–4% CO_2) atmosphere results in a reduced rate of respiratory losses and retards the rate of sucrose reduction compared to normal atmosphere storage. Carrot quality deterioration during storage is due to the loss of sugar in respiration. Atmospheres with less than 10% O_2 and CO_2 at 2.5% reduced carrot rooting and sprouting compared with normal atmosphere storage (Weichmann and Ammerseder, 1974). However, atmosphere conditions of less than 8–10% O_2 and greater than 5% CO_2 were injurious to the roots, and O_2 levels less than 10% tended to increase decay (Apeland and Hoftun, 1969). Ryall and Lipton (1972) suggested that 2–4% CO_2 helped to prevent the development of bitterness. Baumann (1974) reported that root carotene content was decreased with increasing levels of oxygen, this change being attributed to increased respiration rates. Low-pressure storage also was without significant benefit. Irradiation was found to be effective for minimizing storage losses due to sprouting (Salunkhe, 1961). Overall, for all storage situations the combination of low temperatures and high relative humidity is of paramount importance. The age or state of root development (the term 'ripeness' is sometimes used), initial crop quality and freedom from damage or disease are other important factors influencing successful storage results. The loss can be slowed with good handling and storage conditions. Bunched carrots need slightly higher humidity than topped roots, but their storage life is only one-quarter or less than that of topped carrots. Weight loss can be further reduced when carrots are stored in plastic bags. The weight loss in plastic bags is about half that in non-bagged roots.

To meet market standards, carrot roots should be clean, uniform, firm, with bright-coloured smooth surfaces, free of disease and injury, and without bitter or off-flavours.

Storage of other umbellifer root crops

Parsnips are hardy and in some production fields plants are left in the field to overwinter until spring harvest. Roots freeze at –1.7°C, and freeze injury appears as water-soaking of the root core, and tissues in the area of the cambium layer become reddish brown. The external root surface also becomes dark brown. To avoid freezing, the plants are mulched in the field, usually with straw, or field-stored after harvest in insulated clamps or ground pits. However, refrigerated storage is more commonly practised. Storage for 4–6 months is possible at 0°C to –1°C and 98–100% relative humidity. As with carrots, and some other crops, it is important to avoid root exposure to ethylene.

Similarly, **Hamburg parsley** plants are hardy and can be field-stored or dug and stored in clamps or root cellars. After storage and before packing, the roots are cleaned, often washed, and graded for size and quality before marketing. Hamburg parsley can be held in refrigerated stores much like carrots or parsnips, and with the tops removed, can be maintained for several months at 0°C and high relative humidity. Good low-temperature tolerance of **skirret** plants also allows this crop to be overwintered or field-stored. Harvested roots are also held in storages at low temperatures for several months, and occasionally are covered with sand to reduce desiccation. The storage life of **arracacha**, usually at about 5°C, is not particularly long. During storage, roots may develop some sweetness as some starch is converted to sugars. **Celeriac** stores very well, and when held at 0–1°C and relative humidity greater than 95%, quality can be maintained for several months. Celeriac CA storage using low oxygen levels was not beneficial in reducing crop loss, and high CO_2 levels (5–7%) increased decay (Weichmann, 1977). Waxing of trimmed celeriac seldom offers any benefit to keeping quality, but often may increase decay. For favourable market presentation the celeriac should be clean, well-trimmed, firm, about 10–15 cm in diameter with weights between 600–700 g.

LEAFY UMBELLIFER CROP STORAGE

Celery storage

Celery is seldom stored for more than a few weeks in the USA because production from different areas provides most markets with a relatively continuous supply. In other countries celery is stored in order to extend marketing periods or to accommodate surplus production. Storage slows desiccation and deterioration of celery although long-term storage results in some quality loss. With early initial cooling, and continued maintenance at 0°C and 95–99% relative humidity, celery will retain acceptable market condition for 2–3 months. To limit wilting, packaging in ventilated wraps or sleeves is beneficial. Because celery in storage can absorb odours, it is important to avoid shared storage with non-compatible odour-producing products.

Celery can 'grow' in storage as exhibited by the elongation of inner petioles. However, the extension is at the expense of the outer petiole turgor and quality loss. Petioles also exhibit anti-geotropic response if laid horizontally, and therefore the stalks should be placed vertically. Long-term storage will result in some loss of green colour, which is a lesser concern for blanched celery. Celery intended for storage usually is not washed, and only is lightly trimmed until prepared for marketing. Final trimming, and sometimes washing, is performed after removal from storage. Once removed from storage, shelf-life tends to rapidly decline.

In storage investigations, Parsons (1960) found that O_2 levels of less than 5% retarded celery yellowing only slightly, and CO_2 could be damaging, and thus adequate ventilation should be used to avoid accumulation of possible dangerous levels of CO_2 or reductions of O_2 below 1%. According to van den Berg (1981), CA storage at 3% O_2 and 5% CO_2 did reduce decay and loss of green colour when maintained at 0°C, and high relative humidity. The good appearance of celery was maintained longer when ethylene (C_2H_4) was scrubbed from the storage atmosphere. Ryall and Lipton (1972) indicated there was little advantage for holding celery under CA conditions during transit, but if the crop were to be stored for a month, CO_2 levels up to 9% could be accommodated without harm. Smith and Reyes (1988) reported weight losses of less than 10% for celery stored for 10 weeks at 0°C in a controlled atmosphere of 1 or 2% O_2 with either 2 or 4% CO_2 and with C_2H_4 scrubbing. When compared with normal atmosphere storage, less loss of colour or flavour was observed and microbial loss was reduced. Although several other research reports indicated some benefits of CA storage, it is not a commonly performed procedure.

Because shelf-life is relatively short without refrigeration, the product is mostly marketed near the production areas. Clean, uniform, well-trimmed plants enhance marketing attributes.

Parsley storage

Good postharvest quality can be retained for 8–10 weeks at 0°C. High relative humidity is essential to prevent desiccation. Packaging in perforated polyethylene bags is beneficial. Postharvest life is extended with CA storage compared with normal atmosphere storage. Recommended CA storage conditions of 10% O_2 and 10–11% CO_2 at 0°C and 95% relative humidity help to retain green colour and saleability (Apeland, 1971). However, long-term storage, with or without CA conditions is seldom practised.

Storage of other leafy umbellifer crops

Essentially none of the leafy umbellifer crops receive a significant storage period because the thin, often fern-like leaves of cilantro, chervil, Japanese hornwort, dill and similar leafy crops are highly susceptible to rapid desiccation. Even with low temperature and high relative humidity handling, shelf-life seldom can be satisfactorily extended beyond 10–14 days. Florence fennel is an exception if most of the finely divided foliage is trimmed. The fleshy basal petiole portions with appropriate low temperature and high relative humidity conditions can have a satisfactory shelf-life of more than a month.

UMBELLIFER SEED STORAGE

The storage of umbellifer seeds will differ depending upon the ultimate use of the seed, whether they are used for propagation or for food and condiment purposes. Assuming the seeds are well-handled, cleaned and dried to a moisture content appropriate for the species, seeds of most species may be stored satisfactorily for several years with careful management of temperature and relative humidity. There are some exceptions, parsnip perhaps being most notable for its relatively short period of viability.

Clearly more care is given to seeds intended for propagation use in order to retain viability. Accordingly, in some situations controlled atmosphere storage is used, and seed may be stored in moisture-resistant or hermetically sealed containers. However, it is important to minimize possible loss of aromatic and other flavouring compounds from seed that is intended for food and condiment use. Therefore, such products also are often stored in airtight containers that avoid contact with moisture or high temperatures.

8

Umbellifer Utilization and Composition

INTRODUCTION

The utilization of umbellifer vegetables and condiment plants is as diverse as are the crops and their plant parts. Umbellifer vegetables are consumed fresh in the raw or cooked state, as well as processed in a variety of ways and products. Because the overall volume of their consumption is relatively small, few vegetable umbellifers could be considered major world food crops. Nevertheless, their nutrient contributions are significant, and root crops such as carrot, parsnip and arracacha do make appreciable caloric contributions. Although the overall contribution of proteins and fats is generally low the accompanying pro-vitamin A carotenoids, vitamin C, dietary roughage and essential mineral nutrients are important dietary components.

Major attributes of most umbellifer plants are their unique flavour compounds and textural characteristics that make foods more pleasurable. A diverse array of compounds account for the distinctive flavours of the different vegetable umbellifers. Prominent contributors are the volatile terpenoids. These are 10- and 15-carbon compounds (mono- and sesqui-terpenes, respectively). These terpenoids are similar to compounds responsible for the odour and flavour of conifer leaves, citrus peel and mint oils. Interestingly, these are synthesized in the same metabolic pathway as carotenoids.

Flavour compounds occur in leaves, stems and root tissues as well as seeds. It is important to recognize that plant tissues contain many of the same compounds, but these differ in concentration and in relationship to each other. It is such variation that accounts for the many unique detectable differences. The essential oils and various flavour and aromatic compounds contained in plant tissues are used for flavouring many foods and beverages, and as fragrances in non-food products. Several umbellifer crops are often used for garnish and others have ornamental purposes. Furthermore, folk medicinal properties are claimed for almost all edible umbellifers.

ROOT UMBELLIFERS

The major umbellifer crops grown for the consumption of their strongly flavoured and crisp-textured storage roots are carrots, parsnips, celeriac and arracacha. Other umbellifers similarly used are negligible in terms of production volume.

Carrot

Utilization

Carrots are the best recognized and most popular vegetable umbellifer. Their consumption exceeds that of any other umbellifer root crop. Fresh carrots are eaten raw in salads and as a snack food for their sweet, flavourful taste and crunchy texture. Usual methods of cooking fresh carrots include steaming, boiling, baking and stir-frying. Although the storage root is the primary product, young tender foliage occasionally is used as a stir-fried potherb, and in salads in China and Japan (Fig. 8.1). For these purposes, the foliage is produced from high density plantings and harvested at an early growth stage before significant root enlargement.

A recent application of what is called lightly processed technology involves peeling and shaping of root segments to resemble small carrots. The root segments, usually ranging between 3 and 7 cm in length are processed

Fig. 8.1. Bunched carrot tops as an edible fresh market product in Japanese market.

with machines that remove the epidermis and some exterior tissues by light abrasion. The abrasive peeling smoothes surfaces and rounds-off cut ends which results in a product having the appearance of a small carrot. The product is commonly referred to as 'baby' or 'cut and peel' carrots (Fig. 7.11). The consumer convenience for using this product in a variety of fresh and cooked preparations accounts for the popularity of this value-added product. Its production has rapidly expanded and represents a significant portion of the total carrot production in the USA.

In addition to fresh uses, considerable volumes of carrots are processed as canned, frozen, dehydrated, pickled and juice products. The kinds of canned and frozen carrot products are considerable. Small roots may be processed intact. Large roots are diced or cut into slices, wedges, strips and other shapes. They are also puréed for use in infant food preparations. Carrot juice consumption is reported to exceed that of tomato juice in Japan. Dehydrated products include carrot chips, flakes and powder. Thin slices of carrots are deep fried, much like those of potato, for snack food use.

Composition and flavour characteristics of carrot roots

The chemical composition of carrot roots for most cultivars is about 88% water, 7% sugar, 1% fibre, 1% protein, 1% ash and 0.2% fat. Thiamine, riboflavin, niacin, folic acid and vitamin C are also present in appreciable amounts (Table 8.1). Crude fibre in carrot roots consists of cellulose, hemicellulose and lignin, with cellulose being the most abundant. Although reference is sometimes made to starch in carrots (Holland *et al.*, 1992) the starch content of roots is generally close to nil. The anthocyanin content in carrot roots can vary from trace amounts in orange roots to more than 1700 mg kg^{-1} in purple or black carrots (Mazza and Miniati, 1993).

Carotenoids

Carrot roots are a rich source of carotenoids. The carotenoids contained in the edible portion can range from 6000 to more than 54,000 μg $(100\ g)^{-1}$, (60–540 p.p.m.) (Simon and Wolff, 1987). The predominant carotenoids in orange-coloured carrots are β-carotene, α-carotene and γ-carotene (Fig. 8.2).

Carrots are the single most important source of dietary pro-vitamin A carotenoids in the USA, accounting for 30% of the total vitamin A available to consumers (Simon, 1992). Both α- and β-carotene are vitamin A precursors found in carrots. High carotene carrots are used as a raw product for carotene extraction. This 'natural' source of carotene serves as a nutritional amendment and as a food colouring. Poultry producers use carrot extracts to improve animal skin and egg yolk colour.

As previously mentioned, the content and rate of carotene change varies with cultivars, production areas and other environmental factors (Simon *et al.*, 1982). Carotene levels are mainly influenced by genotype. Selective breeding has increased fresh carrot storage root carotene levels of many

Table 8.1. Nutrient composition of several vegetable umbellifers.

Crop	%		g 100 g^{-1} fresh weight					I.U.	mg 100 g^{-1} fresh weight									
	Water	K cal.	CHO	Protein	Fat	Fibre	Ash	Vit. A	Vit. C	B_1	B_2	Niacin	Ca	P	K	Na	Mg	Fe
Arracacha *Arracacia xanthorrhiza*	73	103	24.9	0.96	0.26	0.85	1.30	1,760	23	0.08	0.04	3.4	65	55	240		64	9.5
Asiatic pennywort *Centrella asiatica*	88		6.7	2.0	0.2	1.6	1.6	730	7	0.09			170	32				5.5
Carrot *Daucus carota*	88	140	9.0	1.1	0.2	1.0	0.9	1,100	9	0.35	0.06	0.8	35	40	332	41	19	0.6
Celeriac *Apium graveolens* var. *rapaceum*	88	40	8.4	1.7	0.32	1.28	1.0	16	8	0.04	0.07	0.8	50	107	305	94	15	0.6
Celery (blanched) *Apium graveolens* var. *dulce*	93	18	3.8	0.85	0.15	0.70	1.1	90	8	0.04	0.05	0.6	35	34	308	116	14	0.5
Celery *Apium graveolens* var. *dulce*	94	18	3.8	0.8	0.15	0.65	0.9	184	8	0.04	0.04	0.6	45	30	320	115	17	0.4
Chervil (fresh) *Anthriscus cerefolium*		57	11.5	3.4					9							102		0.5
Chervil (dried leaves) *Anthriscus cerefolium*	7.2	273	49.1	23.2	3.9	11.3	16.6						1,346	450	4,740	83	130	32
Coriander (leaf) *Coriandrum sativum*	88	32	6.0	2.7	0.5	1.0	1.7	10,000	150	0.01	0.01	1.0	150	55	540	28	37	6

Continued

Table 8.1. *Continued*

Crop	%		g 100 g^{-1} fresh weight					I.U.	mg 100 g^{-1} fresh weight									
	Water	K cal.	CHO	Protein	Fat	Fibre	Ash	Vit. A	Vit. C	B_1	B_2	Niacin	Ca	P	K	Na	Mg	Fe
Coriander (seed) *Coriandrum sativum*	11.4		22.7	11.5	19.1	28.4								500				
Florence fennel *Foeniculum vulgare* var. *azoricum*	89	32	6.1	2.4	0.4	0.5	1.7	3,500	30	0.02	0.01	0.4	95	53	390	85	35	2.2
Japanese honewort *Cryptotaenia japonica*	93	18	2.1	2.0	0.1	1.3	1.0	800	60	0.01	0.02	0.5	81	45	490	7		1.9
Parsley (leaf) *Petroselinum crispum*	85	46	8.5	2.8	0.46	1.4	2.0	6,050	123	0.12	0.23	1.1	160	57	508	90	61	4.3
Parsley (dried leaf) *Petroselinum crispum*	8.5	276	51.7	22.4	4.2	10.0	12.2	23,340	120		1.0	8.0	1468	350	3805	437	250	98.0
Parsley (root) *Petroselinum crispum* var. *tuberosum*	88	26	2.3	2.8	0.6		1.62	Trace	41	0.01	0.09	2.0		57				
Parsnip *Pastinaca sativa*	80	79	17	1.7	0.47	2.0	1.2	190	20	0.01	0.01	0.35	50	77	400	11	29	2.0
Smallage *Apium graveolens* var. *secalinum*	91	27	4.6	2.2	0.6	1.4	1.7	2,685	49	0.08	0.12	0.6	326	51	318	151		1.5
Squawroot *Perideridia gairdneri*	60	154	30.8	4.6	1.8				13	0.11	0.12	3.0	440	165	340	12	32	6.5
Water dropwort *Oenanthe javanica*	92	110	4.4	1.1	0.04	1.0	1.5		61		0.31		138	43				2.3

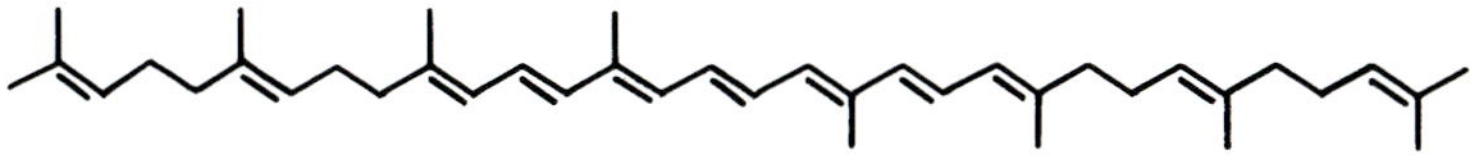

ζ-Carotene

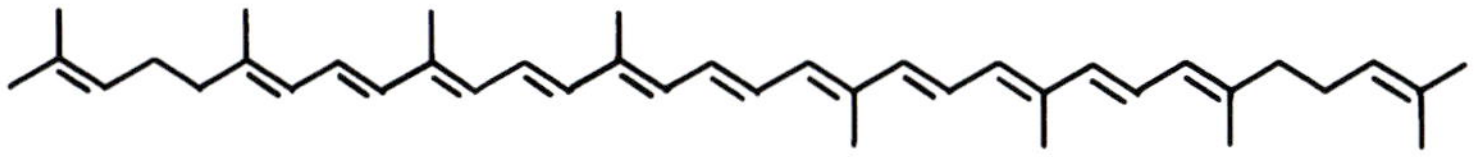

Lycopene

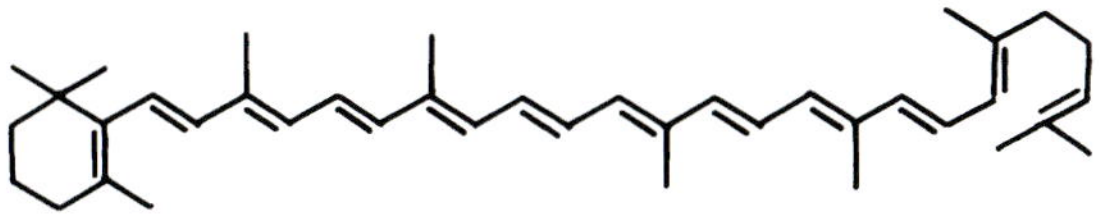

γ-Carotene

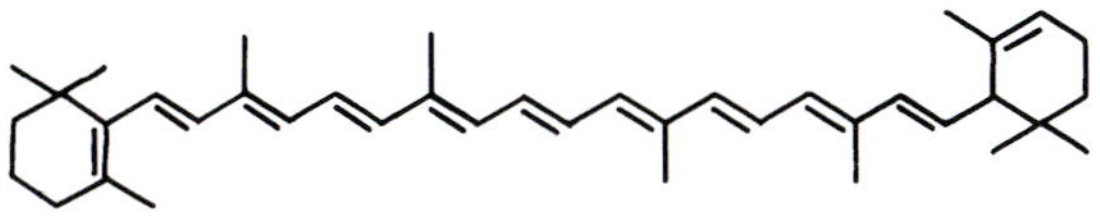

α-Carotene

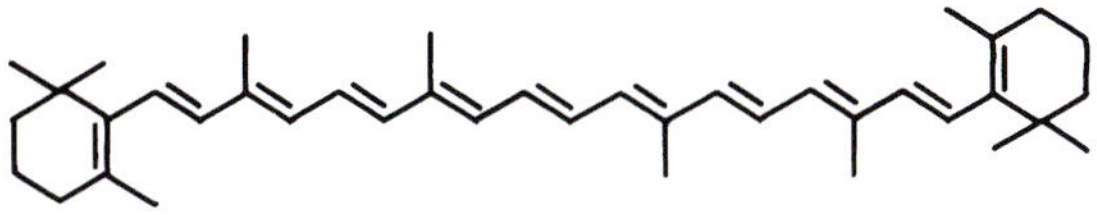

β-Carotene

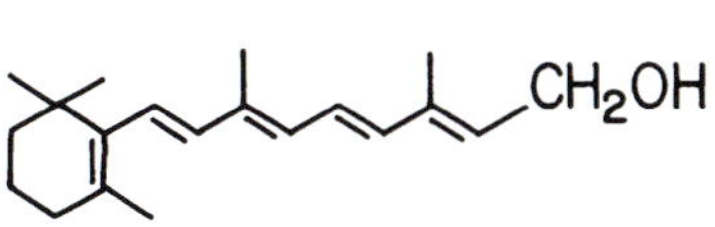

CH_2OH

Retinol

Fig. 8.2. Chemical structure of some carotenoids and retinol (vitamin A).

presently grown cultivars to about 150 p.p.m., and that trend is continuing.

Cooking results in the loss of carotene. Less is lost if carrots are steamed rather than fried, and more is lost as cooking time is increased. During canning an apparent increase of carotene content is observed. However, when leaching of soluble solids was accounted for, it was observed that carotene content had decreased (Edwards and Lee, 1986). Simon and Lindsay (1983) reported the carotenoid level decreased by 7 to 12% for canned cooked carrots of four hybrids tested.

With dehydrated carrots, carotenoid content decreases with length of storage. Drying procedures and storage conditions affect carotenoid stability, sensory changes and volatile constituents.

For stored carrots, the carotene content usually is either maintained or even slightly increased in good storage conditions up to the time sprouting occurs. Root moisture loss may account for some of the observed increase. In some situations the initial carotene content may decrease at the beginning of storage. The length of storage and storage conditions can influence the stability of carotene levels. Levels of specific carotenoids show more variability than changes in total carotenoids.

Phenolic compounds

Phenolic compounds contribute to some organoleptic properties of fresh and processed carrots. These compounds are directly oxidized by reductase enzymes such as polyphenoloxidase to produce bitter compounds like coumarins or indirectly produce tissue-browning following tissue injury. The phenolic compounds appear to have a role in resistance to some fungal and bacterial agents, possibly due to suberization and lignification processes as protection against microbiological attack. The most prevalent phenol is chlorogenic acid. When carrot plants are subjected to the stress of cold, injury and/or ethylene exposure chlorogenic acid accumulates in the roots. Aubert *et al.* (1993) suggested that phenolic compounds of carrots appear as good indicators of cultivar suitability for cold storage as well as for pre-cut product preparations.

Flavour compounds and flavour characteristics

No single compound has been found to account for a distinctively 'carrot-like' flavour. However, it is possible to determine several compounds which contribute to carrot flavour and aroma. Variations in carrot flavour can be attributed to certain genetic, environmental and postharvest factors.

Fresh carrot flavour is primarily attributable to mono- and sesqui-terpenes and sugars (Simon *et al.*, 1980a). Terpinolene is the most abundant monoterpene whereas bisabolenes and caryophyllene are the predominant sesqui-terpenes (Buttery *et al.*, 1968; Heatherbell *et al.*, 1971; Table 8.2).

With very low quantities of these components (<5 p.p.m.), the distinctive fresh carrot flavour may be lacking and other compounds, probably pyrazines, become noticeable. McLellan *et al.* (1983), studying carrot aroma characteristics, reported that the aroma constituents play a minor role in taste parameters for acceptance of fresh carrots.

A wide array of other compounds found in the carrot root also influence

Table 8.2. Predominant flavour compounds of vegetable umbellifers.

Crop	Crop portion(s)	Flavor description	Major flavor components
Anise	Seeds, leaves	Liquorice, sweet taste and smell	Anethole, methylchavicol, limonene, other monoterpenes
Angelica	Leaves, seeds	Musk-like, strongly aromatic juniper-like	Angelicin, methylcamphene, monoterpenes
Asafetida	Oleo-gum resin	Strong garlic-like taste and odour, acrid	Organic sulphur compounds, mercaptans
Caraway	Seeds	Aromatic, sharp, sweet, parsnip-like smell	+/– Carvone, limonene, other monoterpenes
Carrot	Roots, leaves	Sweet, bitter and occasionally turpentine-like taste	Monoterpenes, sequi-terpenes (harsh flavour), sugars (sweet flavour), 2-nonenal (cooked flavour), 3-*sec*-butyl-2-methoxy-pyrazine, eugenin, 3-methyl-6-methoxy-8-hydroxy-3,4-dihyroisocoumarin (bitterness)
Celery	Leaves, seeds	Slightly bitter	Apiole, limonene, 3-butyl-phthalide, sedanolide (celery flavour), non-flavonoid glucoside (bitterness) sesqui-terpenes and other monoterpenes
Coriander	Seeds, leaves	Fragrant, warm, slightly pungent, orange, lemon sage, sour taste of leaves	D-Linalool, limonene, borneol, other monoterpenes, decylaldehyde
Cumin	Seeds	Aromatic, sweet, warm, stinging, spicy lemon flavour, more bitter than caraway	Cuminaldehyde, limonene, pinene, dipentene, other monoterpenes
Dill	Seeds, leaves	Warm, aromatic, faint sweet caraway flavour	+/– Carvone, limonene, alpha-phellandrene, other monoterpenes

Continued

Table 8.2. *Continued*

Crop	Crop portion(s)	Flavour description	Major flavour components
Indian dill	Seeds	Strongly aromatic, caraway flavour	Limonene, +/– carvone, dillapiole, dihyrocarveol, other monoterpenes
Florence fennel	Leaves, seeds	Anise-like, aromatic, sweet liquorice/anise flavour	Anethole, fenchone, limonene, carvone, other monoterpenes
Bitter fennel	Seeds	Anise-like, aromatic, sweet	Anethole, fenchone, other mono-terpenes
Parsley	Leaves	Aromatic, pungent, bitter	Apiole, apiin, apigenin, bergapten, myristicin, other monoterpenes
Parsnip	Roots	Carrot-like, sweet	Apiole, sesqui-terpenes, 3-*sec*-butyl-2-methoxypryazine, other monoterpenes
Japanese hornwort	Leaves	Celery-like	Mesityl oxides, methyl isobutyl, ketones

flavour. Cooked carrot flavour is imparted by 2-nonenal (Buttery *et al.*, 1968). Various phenolic compounds impart bitterness, 3-*sec*-butyl-2-methoxy-pyrazine contributes to carrot aroma (Cronin and Stanton, 1976), amino acids, especially glutamic acid also contribute to sapid components of carrot flavour that are described as 'delicate' (Otsuka and Take, 1969), and ionones impart floral off-flavour (Ayers *et al.*, 1964). Texture also is an important aspect of sensory quality.

Sweetness

It is well-recognized that high sugar levels are generally preferred by carrot consumers and that sweetness in carrots is primarily due to the sugars: glucose, fructose and sucrose. Reducing sugars accumulate early in carrot root development while sucrose usually becomes predominant by harvest time. These sugars are stored in carrots at levels that vary during root development and among cultivars from 4% up to 10% of fresh weight. Simon and Lindsay (1983) reported that reducing sugars accounted for 6–32% of the free sugars in several hybrid carrot cultivars and it is also known that high solids improve sweet flavour. A single gene controlling the balance of reducing sugars to sucrose has some influence on flavour (Freeman and Simon, 1983). Accordingly, many carrot breeding programmes select for high solids levels in carrots since such selection can be effective for improving sweetness and flavour (Stommel and Simon, 1989).

Harshness

The harsh flavour of foliage and fresh storage roots is due to volatile mono- and sesqui-terpenoids, and is especially noticeable when sugar content is low. As root sugar content increases, the taste of the terpenoids is masked to some extent, but a balanced, relatively low volatile terpenoid level is necessary for optimal fresh carrot flavour.

When the cumulative mono- and sesqui-terpene quantity exceeds 30–50 p.p.m. on a fresh weight basis, a harsh turpentine-like flavour is noted. Lower levels do not elicit this response, but do contribute to the typical fresh carrot flavour. Flavour terpenoids occur in oil ducts at levels of 5–500 p.p.m. fresh weight (Senalik and Simon, 1987).

The undesirable harsh or bitter flavours cannot be totally masked by high sugar content. Astringency and turpentine-like taste is perceived more in the throat than the mouth.

Bitterness

Much research has been directed to the understanding of bitter flavour development in fresh and stored carrots. Ethylene triggers the development of bitter flavour which is readily detected in both fresh and processed carrots, and even when carrots are only a minor component in a mixture with other vegetables. Stored carrots do not generate enough ethylene to stimulate bitter flavour development, but just a few days of carrot storage in the presence of ethylene-producing fruits, such as apples or pears, can ruin quality. The bitter-tasting compound is reported to be isocoumarin (3-methyl-6-methoxy-8-hydroxy-3,4-dihydroisocoumarin) (Sondheimer, 1957). When ethylene sources are avoided, carrot flavour can remain relatively unchanged for 5–6 months of storage at 0–2°C and high relative humidity. It is not known whether the bitter flavour developed in aster yellows-infected carrots is due to the same compound as that induced by ethylene.

Effects of processing

Fresh cooked carrots generally had slightly more sugar than raw carrots but canned-cooked and frozen-cooked samples lost 20–45% of their total sugar (Simon and Lindsay, 1983). These types of processing also hydrolysed some of the sucrose to reducing sugars. The extent of hydrolysis varied between genotypes, with little effect upon those containing high reducing sugar (30% of total sugar), but doubling reducing sugar levels from 7 to 15% in other genotypes.

Processing reduced volatile terpenoid content by 50–85%, with less loss after fresh cooking than with canned-cooked, frozen-cooked or freeze-dried carrots (Heatherbell *et al.*, 1971; Simon and Lindsay, 1983). Most of the reduction is due to the loss of the lower boiling mono-terpenes. With volatile terpenoid loss in processing, the harsh flavour is diminished. The decrease is larger in canned- or frozen-cooked products than fresh-cooked carrots. Compounds increasing with canned-cooked carrots are the aldehydes, namely 2-nonenal which has a soft sweet flavour. The increase is very small, but the detection threshold for this compound is very low and it is very evident and important in cooked carrot aroma.

When freeze-dried carrot is stored in the presence of oxygen, colour is lost due to destruction of carotenes and violet-like off-flavour aromatic compounds (ionones) are produced, presumably from oxidation of the carotenoids (Ayers *et al.*, 1964). Freeze-drying causes the loss of as much as 75% of the volatiles contained in the essential oil. Mature carrot roots stored for 5 weeks accumulated large quantities of acetaldehyde and ethanol which is indicative of anaerobic respiration. Concentration of other volatiles did not change significantly, and flavour changes were not noted. Bitter flavour is not noticeably affected by processing. However, processing of briefly stored carrots results in less bitterness than those from a long refrigerated storage.

Parsnip

Parsnips possibly are the second most well-known umbellifer root crop. Their popularity is highest in northern European countries. The roots are usually prepared for consumption by steaming or roasting, and are more often used as an ingredient in stews and other cooked meals rather than as a separate vegetable dish. Parsnips are processed by canning, although that use is much less than for fresh use. Parsnips accumulate starch as a storage carbohydrate. Sweetness is generally found to increase in parsnip roots very late in the growth period when temperatures typically decline. Frost and low-temperature exposure tends to convert some of the root starch into sugars which produces a noticeably sweeter taste. For this reason, freshly harvested roots are often stored for a brief period at low temperatures to increase the sweet taste before marketing. Parsnip flavour is due to a group of compounds similar to those eliciting carrot flavour such as mono- and sesqui-terpenoids, sugars and 3-*sec*-butyl-2-methoxy-pyrazine (Johnson *et al.*, 1971).

Celeriac

Much like parsnip, celeriac is a relatively popular commodity in Europe (Fig. 8.3). Most often celeriac is cooked, but is also eaten raw in salads. Peeled segments are frequently added to stews and soups for bulk and flavouring. A sizeable volume is processed by canning and pickling. Although edible, celeriac foliage is much too coarse and generally is not used for food, but does find use as a flavouring bouquet. **Hamburg (turnip-rooted) parsley** and **turnip-rooted chervil** are other vegetables whose popularity is mostly European. The cooked enlarged storage roots of Hamburg parsley are used in soups and as a boiled vegetable, and are especially popular during holiday

Fig. 8.3. Celeriac roots trimmed for fresh market presentation.

periods in eastern Europe where the crop is known as 'petrouska'. Similarly used are the spindle-shaped sweet-tasting storage roots of turnip-rooted chervil. Secondary in importance to the roots, the fresh leaves of turnip-rooted chervil provide a flavour which resembles salad chervil, although it is a different species.

Arracacha

In contrast with most other root umbellifer crops, arracacha is an important New World crop. In some South and Central America locations, arracacha represents a fairly significant source of calories. The roots resemble carrot and parsnip in appearance, with a flavour that also resembles those of carrot, parsnip and celery.

Arracacha roots are consumed after cooking in a multitude of preparations that include boiling, baking, frying, for use as whole, mashed or puréed vegetables in soups, and stews. Fried chips, which have a crispness that some feel is superior to that of potato, also are a popular product. Processed products include dehydrated powder which is widely used for instant food preparations, mostly for soups, baby foods, and various pastry and bakery products. Starch is another product. The dry weight starch content of the roots ranges from 10 to 25%, and the small starch granules being highly digestible are an excellent food for infants and the elderly. Arracacha roots are also used in preparation of mild alcoholic beverages. One in particular is known as 'chicha'. Arracacha foliage has similar characteristics to those of celery. When intended for vegetable use, the leaves and petioles are blanched in order to reduce bitterness and enhance tissue succulence.

Additional umbellifer root crops

After cooking, the white-fleshed roots of **skirret** have a taste that closely resembles that of salsify roots. Starchy storage roots of **epos**, which were a food source for the indigenous people and early settlers in North America, can be eaten raw and cooked. Additionally, the roots can be dried and stored for winter use. The large roots of **lovage** and **angelica** after peeling are boiled and used as a vegetable. However, more often they are used for the flavouring that is provided by the essential oils extracted from the roots, and by powders prepared from ground dried roots. These have a broad range of flavouring uses for food and non-edible products such as perfumes, soaps and tobacco. Some medicinal uses are also made of the dried roots and root extracts. The resinous exudate from **asafetida** roots has a strong bitter, sharp, stinging, pungent onion and garlic-like taste and odour due to sulphur compounds. The gum-like extract is a very popular addition in Indian and Iranian food preparations, and is an ingredient in soups, Worcestershire sauce and ice-cream.

The enlarged storage roots of some lesser-known umbellifers consumed as vegetables include: **burnet saxifrage**, **cow parsnip** and **great earthnut**. Roots of several umbellifers not generally considered root crops are also occasionally consumed as vegetables. Some examples are: **sweet cicely**, **caraway**, **lovage** and **coriander**. The latter is a fairly popular root vegetable in some areas of China. Root oil extracted from burnet saxifrage is used in flavouring candies, liquors and some pharmaceuticals.

FOLIAGE UMBELLIFERS

The primary attributes of the leafy vegetable umbellifers are the contributions of flavour, vitamins, other nutrients and dietary fibre that their consumption provides. As commonly utilized, the leaves and the thickened petioles and stem tissues do not contribute significant bulk or calories.

Celery

Celery is a major leafy vegetable having many food uses. Its succulent petioles are eaten raw as cut sticks and slices in salads, and as a cooked potherb. Celery is also extensively used for its flavour and textural contributions to soups, stews and other meal preparations. Some celery and smallage cultivars, known as 'cooking celery', are used specifically for their very intense flavouring qualities.

The distinctive flavour of celery petioles and seeds is due to the presence of phthalides and terpenes. Celeriac and smallage tissues also contain these compounds. Other compounds that celery and some other umbellifers contain are linear furanocoumarins, such as psoralen, xanthotoxin, bergapten and isopimpinellin in their foliage. The first three compounds are phototoxic, causing dermatitis in human and animals after contacting the skin or being ingested and followed by exposure to sunlight. Some individuals exhibit much greater sensitivity to psoralens than others. Normally the concentration of these compounds in celery, parsley and several other umbellifers does not pose a health threat for consumption or to field workers handling these plants. The concentrations of these compounds have been found to increase in response to pollutants, cold temperature, fungal infections, mechanical damage and the ultraviolet spectrum of sunlight.

Celery petioles are processed by canning, pickling, dehydration and juicing. Typically, the juice is combined with other vegetable juices for the final product. Juice and seed oil are also used to flavour table salt, some non-alcoholic beverages and as an ingredient in fragrance products. The volume of celery processed by freezing is limited because the crisp textural characteristics deteriorate when the product is thawed. However, this is acceptable when celery is an ingredient of vegetables mixes or other food preparations. Celery seeds are a common and frequently used condiment as are dried leaf blades in the form of flakes and powders. Seeds are used directly and seed oils are also extensively used in many flavouring practices. A large volume of celery seed is actually produced from crops of smallage. Bitterness in fresh and stored celery has not been as well-characterized as that in carrot, but is associated with the presence of a non-flavonoid glucoside (Pan, 1961).

Overall, the production and utilization of **smallage** greatly surpasses that of celery. Having broad climatic adaptation permits smallage production in many locations. Accordingly, smallage is a very important fresh and cooked vegetable for salad, potherb and food-flavouring purposes for the large populations of Asia and other countries. Considerable amounts are processed by dehydration.

Parsley also is a high-volume leafy vegetable utilized for its aromatic, tangy sweet-tasting leaves. Used raw or cooked, parsley adds flavour to many food preparations. In general the flavour is too intense to be consumed as an unaccompanied potherb. Some of parsley's popularity is also due to its high β-carotene and vitamin C content. Leaf carotene contents can range from 30 to as much as 60 p.p.m. In addition to food uses, parsley foliage is widely used as a garnish. It is generally recognized that the flat leaf parsley cultivars are more flavourful than the curly leaf types, the latter more often are used for garnishing meals. The dehydrated foliage is prepared as flakes, and extensively used for flavouring.

Florence fennel is a crop greatly appreciated in some regions, namely southern Europe, but otherwise is of limited popularity. The bulb-like petiole bases are used fresh in salads for their crisp texture and sweet strongly flavoured anise/liquorice-like taste. The crop is also a popular cooked potherb and is often prepared in pickled form. The finely cut foliage is occasionally used for seasoning and as a garnish.

Cilantro, also widely known as Chinese or Mexican parsley, and more often simply as coriander, is a popular leafy crop that provides a unique flavour and aroma to numerous foods (Fig. 8.4). Cilantro is almost indispensable to daily meal preparations in India, China and countries in South-east Asia, the Near East, the Mediterranean basin and Central America. Recently the consumption of this crop has greatly expanded in Europe, the USA and Canada. The fresh leaves have a taste that resembles a blending of lemon and parsley. The predominating odour of fresh cilantro is due to the aldehyde content of its essential oils. The odour, due to decylaldehyde and other fatty aldehydes, is unpleasant to some, but that usually is compensated by the slightly pungent flavour the leaves contribute to foods. Cilantro shares many of the uses of parsley, and its aromatic flavouring qualities are used in numerous culinary preparations. These include fresh use in salads and for soups, stews and other

Fig. 8.4. Bunched cilantro foliage.

meal preparations. It is also used in chutneys, salsa and various other sauces. Both fresh and dried fresh leaves are popular for flavouring fish and meat preparations. Old leaves and leaves harvested after plants begin flowering should be avoided as these lose their flavour qualities and palatability.

Culantro, used very much like cilantro, is very popular in Central America, the West Indies and countries of northern South America, as well as South-east Asia, where one of its most common uses is for flavouring rice dishes. The petioles of **Japanese hornwort** are aromatic and have a pungent parsley-like odour and taste (Fig. 7.7). The young leaves and tender petioles are eaten raw in salads, used for flavouring of many different foods and cooked as greens. The petioles are frequently blanched in order to increase tenderness. Roots are also eaten. The leaves and petioles of **water dropwort** have similar uses, and commonly are used raw much like celery, and also are cooked, usually by steaming or stir-frying.

Chervil has many admirers, especially in France and several other European countries, due to its delicate unique flavour. Both of the two leaf types grown (flat and curled) are more aromatic than parsley, and have an anise-like fragrance. The taste is slightly pungent or peppery and resembles a combination of parsley and anise. Leaves are frequently used in salads and to flavour egg dishes, other vegetables, fish and cheese. Chervil is used in many herbal seasonings, i.e. 'fines herbes' and 'bouquets garnis', and as a garnish.

Lovage leaf petioles and stem portions contribute a slightly pungent celery-like taste to salads, soups and stews. When produced for vegetable use the plants are blanched during growth. A well-known use is candied peeled tender stems. **Black lovage** young shoots and leafstalks also have a pungent celery-like taste, and are used fresh as a potherb. Plants also are blanched during growth to improve textural characteristics. The young leaves of **sweet cicely** have a sweet taste much like anise and are used for flavouring soups and stews, and also used fresh in salads and as garnish.

The aromatic pungent leaves and tender stems of **lawn pennywort** and **Asiatic pennywort** are eaten uncooked, as a potherb, or as a flavour accompaniment, most often with steamed rice or porridge, and also as an ingredient in various sauces. Fresh extracted juice is also used in various food preparations. These vegetables have a flavour and fragrance that resembles parsley. Both lawn and Asiatic pennywort are very popular in Indonesian and Southeast Asian meals.

Angelica leaves have a musky odour and sweet taste. Young tender stems and leaf petioles are peeled before eating fresh in salads or as a potherb, and also are added to other food preparations for flavouring. Stems, tender shoots and petioles are often prepared in candied form. Fresh leaves are used directly or as a dried ground powder for many flavouring purposes. Extracted oil from stems is used for flavouring ice-cream, candy, baked goods, gin, vermouth, other kinds of liquors as well as toothpaste, cigarettes and perfume. Other species of angelica are also similarly utilized, more so for their flavouring rather than edible quality characteristics.

Dill weed is the term commonly used to refer to the harvested foliage of dill plants. The foliage, in fresh and dried form, has multiple seasoning and flavouring uses for salads, soups, sauces and salad dressings (Fig. 8.5). Leaves

Fig. 8.5. Bunched dill foliage.

and dill seeds are also used to flavour pickles, meats, fish and breads. Dill foliage and dill seeds differ in taste because their respective tissues have a different composition of flavouring compounds. The seeds tend to have more of a caraway taste than does the foliage. **Indian dill** foliage is widely used in many Asian countries to flavour rice and soups, and also has uses similar to dill weed and dill seed. **Anise** fresh and dried leaves are used in salads, soups and stews, and for garnish. These aromatic tissues are widely known for their flavour represented by the powerful, sweet, aromatic liquorice-like taste.

Fresh leaves and shoots of **burnet saxifrage** are used in salads. They have a noticeable mild cucumber odour. Dried foliage is used in the flavouring of beer and wine. In some production areas the edible foliage of **asafetida** plants are used as a potherb. The cabbage-like terminal vegetation is the plant portion most preferred. Largely confined to European countries, the leaves and stems of **Bishop's weed** are used as a potherb and fresh in salads. Its aromatic foliage has a taste most like spinach, but with a tangy aspect. The strongly scented succulent young shoots and leaves of **samphire** are used fresh in salads or as cooked potherbs, and seed pods are pickled.

Although minor in terms of usage, leafy portions of several other umbellifers are occasionally utilized as potherbs, for salads and as flavouring ingredients. Some better known examples are turnip-rooted chervil, Hamburg parsley, caraway and ajowan.

CONDIMENT UMBELLIFERS

Several umbellifer plants are prized as condiments or savoury herbs because of the essential oils and various other flavour compounds contained in their tissues. Even in small amounts these ingredients provide unique flavours, individually and when accompanying various prepared foods. Often it is the seed that contains the highest concentration of flavouring agents for condiment plants.

Coriander seeds and seed oil have extensive worldwide flavouring uses, and in terms of volume is a world leader. When immature, coriander seeds have a tangy citrus and sage-like taste, when mature the seeds are pungent. Coriander is an important component of many curry powders, often comprising one-third of all the other ingredients. The seeds are used as pickling spices, to flavour sausages, other meats and fish.

Seed essential oils are distilled and have many flavouring uses in candy, and cocoa and chocolate production. Other uses include flavouring of alcoholic beverages (gin and vermouth) and tobacco, in addition to use in the manufacture of some perfumes and colognes. Seed-extracted fatty oils have various industrial uses for their physico-chemical properties. Different coriander plant types have been developed which are better suited for seed production, whereas others are preferred for foliage (cilantro) production.

Caraway seeds and seed oils are multiple-purpose flavouring products extensively used for flavouring cookies, cakes, cheese, sauerkraut and rye bread, in addition to use in the preparation of candies, seasoning meats, pickles, beverages and certain well-known liquors such as kümmel and schnapps. The seeds have a spicy and sharp taste.

Seeds of **bitter** and **sweet fennel** are used to flavour cookies, baked goods, cheese, and various meat and vegetable dishes. Fennel seeds are less pungent than those of dill. Fennel seed oils are also used for aromatic fragrance production. Both sweet and bitter fennel seeds contain similar levels of anethole, but differ in that bitter fennel contains more fenchone (Betts, 1968). Bitter and sweet fennel produce a higher yield of larger and more flavourful seeds than those of Florence fennel.

Cumin seeds are an important ingredient in curries and chilli powder preparations, and for flavouring sausages. Other uses include flavouring pickles, chutneys, breads, cakes, cheese, sausages, other meats, some liquors and perfumes. The seeds are aromatic and spicy, but more bitter than those of caraway. **Dill** seeds have multiple seasoning and flavouring uses. They are widely used in salads, soups, sauces and dressings, and for flavouring pickles, meats, fish and bread. Dill seed oil is an important ingredient for candies and gum. Dill weed oil is extracted from the foliage, usually by distillation, and differs in composition and has some different uses to the oil extracted from crushed seeds. The pungent and somewhat bitter seeds of **Indian dill** also are an important component in many curry powders, and some pastries and beverages, especially in India and Indonesia. The composition of Indian dill seed oil differs from dill seed oil, mostly because it has a lower carvone content.

All parts of the **anise** plant are fragrant. The seeds are the major product and are used directly, and as extracted seed oil for flavouring meat products, sausages, baked goods, pasties, cheese and beverages. A liquor known as anisette, some wines, candies, gum, ice-cream, cough syrups, mouthwashes, toothpaste, pickles, tobacco and scents are other products that utilize anise.

Lovage seeds resemble caraway and have some similar uses in candies, cakes and meat dishes, or are pickled for use much like capers. Seed oil extracts are often added to pickling mixes, confectioneries, syrups and for flavouring cheese, sauces and some alcoholic beverages. **Sweet cicely** seeds are used in cakes and baked products, and the spicy seed oils for flavouring candies, some beverages and in particular a liquor known as Chartreuse. Extracted seed oils of **angelica** are also used for flavouring candies and various confectionery products, alcoholic beverages that include gin and vermouth, as well as toothpaste, cigarettes and perfumes. A primary use of the aromatic pungent seeds of **ajowan** is in curries, but the extracted seed oil which has an odour resembling thyme is also used for flavouring purposes. Oleoresins from seeds of other species of *Ferula* are also used by the flavouring industry.

MEDICINAL ATTRIBUTES

In addition to their caloric, fibre and nutrient contribution to the diet, several benefits to human health and comfort are ascribed to umbellifer plants. Apiaceous (umbelliferous) plants have long been consumed for their perceived medicinal qualities. For example, before AD 900, it appears that carrot seeds were utilized in preference to the storage roots. That situation was similar for other umbellifer plants.

Most of the putative medicinal attributes of the different umbellifers are similar or repetitious. In general, they are said to provide relief for flatulence, fever, rheumatism, menstrual discomfort, halitosis, and provide diuretic, expectorant and anti-spasmodic properties, as well as acting as an appetite stimulant, digestive aid and several other benefits.

It is interesting to note that the umbellifers also have some history of toxicogenic qualities ranging from the death of Socrates (due to consumption of alkaloid-containing *Conium maculatum*) to the occurrence of dermatitis-causing psoralens in the foliage of many umbellifers.

The carotenoids, β-carotene in particular, are well-known for their anti-mutagenic, fungal and bacterial properties, as well as chemopreventive, photoprotective and immuno-enhancing properties. These effects are mainly linked to the antioxidant properties of these compounds that may account for the biological and medicinal properties of carrot and some other umbellifers. β-Carotene is helpful for relief of certain skin conditions, and reducing the risk of cataract formation. Other reported benefits of carrots and other umbellifer crops, such as reducing the risk of several types of cancer, require further validation.

NEW AND FUTURE DIRECTIONS

With the wide array of carotenoid pigments and other compounds thought to impart health benefits in carrots, the ease of producing large biomass of carrot roots, and the ease of genetic manipulation of carrot with classical and biotechnological methods makes carrots a prime target in the development of 'functional foods'. These are foods with particular enhanced health benefits. The feasibility of producing functional foods on a large scale remains to be tested. It would be interesting to examine the possibilities for identifying sources of yellow- or orange-rooted parsnips, root parsley or orange petioles in celery. Such germplasm sources exist for arracacha storage roots, but could be enhanced. Alternatively, it may be feasible to use transgenic technologies to insert key genes controlling the carotenoid pathway into these crops to yield higher carotene levels. Flavour profiles may also be able to be significantly altered in vegetable umbellifers with genetic and biotechnological manipulation.

SELECTED REFERENCES

Aloni, B. and Pressman, E. (1979) Petiole pithiness in celery leaves: induction by environmental stresses and the involvement of abscisic acid. *Physiologia Plantarum* 47, 61–65.

Aloni, B., Bar-Yosef, B., Sagiv, B. and Pressman, E. (1983) The effect of nutrient solution composition and transient nutrient starvation on the yield and petiole pithiness of celery plants. *Journal of Horticultural Science* 58, 91–96.

Angell, F.F. and Gabelman, W.H. (1968) Inheritance of resistance in carrot, *Daucus carota* var. *sativa*, to the leafspot fungus, *Cercospora carotae*. *Journal of the American Society for Horticultural Science* 93, 434–437.

Apeland, J. (1971) Factors affecting respiration and colour during storage of parsley. *Acta Horticulturae* 20, 43–52.

Apeland, J. and Hoftun, J. (1969) Physiological effects of oxygen on carrots in storage. *Acta Horticulturae* 20, 108–114.

Arold, G. (1987) Cultivation experiments with parsnips [Anbauversuche zu pastinaken]. *Gemuse* 23(5), 254–256.

Arus, P. and Orton, T.J. (1984) Inheritance patterns and linkage relationships of eight genes in celery. *Journal of Heredity* 75, 11–14.

Atherton, J.G., Basher, E.A. and Brewster, J.L. (1984) The effects of photoperiod on flowering in carrot. *Journal of Horticultural Science* 59(2), 213–215.

Atherton, J.G., Craigon, J. and Basher, E.A. (1990) Flowering and bolting in carrot. I. Juvenility, cardinal temperatures and thermal time for vernalization. *Journal of Horticultural Science* 65(4), 423–429.

Aubert, S., Bonnet, A. and Szot, B. (1979) Texture and quality of carrot. *Annals of Technology and Agriculture* 28, 397–422.

Aubert, S., Babic, I., Amiot, M.J. and Nguyen-The, C. (1993) Les composés phénoliques marqueurs de la qualité de carottes conservées au froid. *Acta Horticulturae* 354, 201–214.

Ayers, J.E., Fishwick, M.J., Land, D.G. and Swain, T. (1964) Off-flavor of dehydrated carrot stored in oxygen. *Nature* 203, 81–82.

Banga, O. (1963) *Main Types of the Western Carotene Carrot and their Origin.* N.V. Uitgevers-Maatschappij W.E.J. Tjeenk Willink, Publisher, Zwolle, The Netherlands.

Baumann, H. (1974) Preservation of carrot quality under various storage conditions. *Acta Horticulturae* 38, 327–337.

Benjamin, L.R. (1982) Some effects of differing times of seedling emergence, population density and seed size on root-size variation in carrot populations. *Journal of Agricultural Science (Cambridge)* 98, 537–545.

Benjamin, L.R. and Wren, M.J. (1980) Root development and source–sink relation in carrot *Daucus carota* L. II. Effects of root pruning on carbon assimilation and the partitioning of assimilates. *Journal of Experimental Botany* 31, 1139–1146.

Benard, D. and Punja, Z.K. (1995) Role of *Pythium* species in cavity spot development on carrot in British Columbia. *Canadian Journal of Plant Pathology* 17, 31–45.

Betts, T.J. (1968) Anethole and fenchone in developing fruits of *Foeniculum vulgare* Mill. *Journal of Pharmaceutics and Pharmacology* 20, 469–472.

Bianco, V.V. (1990) Finocchio. In: Bianco, V.V. and Pimpini, F. (eds) *Orticultura.* Patron Editore, Bologna, pp. 168–183.

Block, A., Dangle, J.L., Hahlbrock, K. and Schulze-Lefert, P. (1990) Functional borders, genetic fine structure and distance requirements of *cis* elements mediating light responsiveness of the parsley chalcone synthase promoter. *Proceedings of the National Academy of Sciences of the USA* 87, 5387.

Boiteux, L., Della Vecchia, P. and Reifschieder, F. (1993) Heritability estimate for resistance to *Alternaria dauci* in carrot. *Plant Breeding* 110, 165–167.

Bonnet, A. (1983) Source of resistance to powdery mildew for breeding cultivated carrots. *Agronomie* 3, 33–37.

Bonnet, A. (1985) Breeding of carrot male sterile lines and F_1 and three-way hybrids creation. *Eucarpia – Breeding of Root Crops*, 30–37.

Borthwick, H.A. (1931) Development of the macrogametophyte and embryo of *Daucus carota. Botanical Gazette* 92, 23–44.

Borthwick, H.A., Phillips, M. and Robbins, W.W. (1931) Floral development in *Daucus carota. American Journal of Botany* 18, 784–786.

Bouwkamp, J. C. and Honma, S. (1970) Vernalization response in celery. *Journal of Heredity* 61, 115.

Bradford, K.J. (1986) Manipulation of seed water relations via osmotic priming to improve germination under stress conditions. *HortScience* 21, 1105–1112.

Bradley, G.A. and Smittle, D.A. (1965) Carrot quality as affected by variety, planting and harvest dates. *Proceedings of the American Society for Horticultural Science* 86, 397–405.

Bradley, G.A., Smittle, D.A, Kattan, A.A. and Sistrunk, W.A. (1967) Planting date, irrigation, harvest sequence and varietal effects on carrot yields and quality. *Proceedings of the American Society for Horticultural Science* 90, 223–234.

Brocklehurst, P.A. and Dearman, J. (1983) Interaction between seed priming treatments and nine seed lots of carrot, celery and onion. I. Laboratory germination. *Annals of Applied Biology* 102, 577–584.

Brocklehurst, P.A., Rankin, W.E.F. and Thomas, T.H. (1982) Stimulation of celery seed germination and seedling growth with combined ethephon, gibberellin and polyethylene glycol seed treatments. *Plant Growth Regulation* 1, 195–202.

Browers, M.A. and Orton, T.J. (1986) Celery. In: Banjai, Y.P. (ed.) *Biotechnology in Agriculture and Forestry*, Vol. 2. Springer-Verlag, Berlin.

Buishhand, J.G. and Gabelman, W.H. (1979) Investigations on the inheritance of color and carotenoid content in phloem and xylem of carrot roots (*Daucus carota* L.). *Euphytica* 28, 611–632.

Burdine, H.W. (1973) The development of pencil stripe in celery. II. Internal induction. A manifestation of brown checking. *Proceedings of the Soil and Crop Science Society of Florida* 32, 77–83.

Buttery, R.G., Seifert, R.M., Guadagni, D.G., Black, D.R. and Ling, L.C. (1968) Characterization of some volatile constituents of carrots. *Journal of Agricultural Food Chemistry* 16, 1009–1015.

Catlin, D., Ochoa, O., McCormick, S. and Quiros, C.F. (1988) Celery transformation by *Agrobacterium tumefaciens*: cytological and genetic analysis of transgenic plants. *Plant Cell Reports* 7, 100–103.

Cerkauskas, R.F. and McGarvey, B.D. (1988) Fungicidal control of phoma canker of parsnip. *Canadian Journal of Plant Pathology* 10, 252–258.

Channon, A.G. (1964) Studies on parsnip canker. III. The effect of sowing dates and spacing on canker development. *Annals of Applied Biology* 54, 63–70.

Channon, A.G., Dowker, B.D. and Holland, H. (1970) The breeding of Avonresister, a canker-resistant parsnip. *Journal of Horticultural Science* 45, 249–256.

Chalutz, E., Devay, J.E. and Maxie, E.C. (1969) Ethylene-induced isocoumarin formation in carrot root tissue. *Plant Physiology* 44, 235–241.

Chessin, D.A. and Hicks, J.R. (1987) The effect of nitrogen fertilizer, herbicides and cultivar on nitrogen components of carrot roots. *Scientia Horticulturae* 33, 67–73.

Cole, R.A. (1985) Relationship between the concentration of chlorogenic acid in carrot roots and the incidence of carrot fly larval damage. *Annals of Applied Biology* 106, 211–217.

Cole, R.A., Phelps, K., Ellis, P.R. and Hardman, J.A. (1987) The effects of time of sowing and harvest on carrot biochemistry and the resistance of carrots to carrot fly. *Annals of Applied Biology* 110, 135–143.

Collier, R.H. and Finch, S. (1996) Field and laboratory studies on the effects of temperature on the development of the carrot fly (*Psila rosae* F.). *Annals of Applied Biology* 128, 1–11.

Constance, L. (1971) History of the classification of Umbelliferae (Apiaceae). In: Heywood, V.H. (ed.) *The Biology and Chemistry of the Umbelliferae.* Academic Press, New York, pp. 1–12.

Constance, L., Chuang, T.-I. and Bell, C.R. (1976) Chromosome numbers in Umbelliferae V. *American Journal of Botany* 63, 608–625.

Cox, E.F. and Dearman, A.S. (1978) The control of blackheart of celery with calcium sprays. *Experimental Horticulture* 30, 1–6.

Craigon, J., Atherton, J.G. and Basher, E.A. (1990) Flowering and bolting in carrot. II. Prediction in growth room, glasshouse and field environments. *Journal of Horticultural Science* 65(5), 547–554.

Cronin, D.A. and Stanton, P. (1976) 2-Methoxy-3-*sec*-butylpyrazine – an important contributor to carrot aroma. *Journal of Science and Food Agriculture* 27, 145–151.

Crowden, R.K. (1969) Chemosystematics of the Umbelliferae, A general survey. *Phytochemistry* 8, 1963–1984.

Cserni, I., Prohaszka, K. and Patocs, I. (1989) The effect of different N-doses on changes in the nitrate, sugar and carotene contents of carrots. *Acta Agronomica Hungarica* 38, 241–248.

Curtis, D.S. (1938) Determination of stringiness in celery. *Cornell University*

Agricultural Experimental Station Memoir 212.

D'Antonio, V. and Quiros, C.F. (1987) Viability of celery pollen after collection and storage. *HortScience* 22, 717–722.

Dame, A., Bielau, M., Stein, M. and Weit, E. (1989) Untersuchungen zum Ertrag an Saatgut bei männlick sterilen Linien der Speisemöhren, *Daucus carota* L. ssp. *sativa* (Hoffm.) Schübl. et Mart. *Archiv für Züchtungsforsch* 2, 111–118.

Dawson, P. (1993) Latest breakthroughs in veg varieties, *Grower*, 9 Sept., pp. 12–13, quoted in article by Edward Long.

Demyttenaere, P., Hofman, G. and Vulsteke, G. (1990) The influence of available nitrate nitrogen in the soil profile on the nitrate contents of blanching celery. In: van Beusichem, M.L. (ed.) *Plant Nutrition – Physiology and Applications.* Proc. 11th International Plant Nutrition Colloquium, Wageningen, The Netherlands, 30 July–4 August 1989. Kluwer Academic Publishers, pp. 747–751.

Diawara, M.M., Trumble, J.T., Quiros, C.F., White, K.K. and Adams, C. (1994) Plant age and seasonal variations in geneotype resistance of celery to beet armyworm. *Journal of Economic Entomology* 87, 514–522.

Dickson, M.H. (1966) The inheritance of longitudinal cracking in carrots. *Euphytica* 15, 99–101.

Dickson, M.H. and Peterson, C.E. (1958) Hastening greenhouse seed production for carrot breeding. *Proceedings of the American Society for Horticultural Science* 71, 412–415.

Diederichsen, A. (1996) Coriander *Coriandrum sativum* L. IPGRI, Rome.

Edwards, C.G. and Lee, C.Y. (1986) Measurement of provitamin A carotenoids in fresh and canned carrots and peas. *Journal of Food Science* 51, 534–535.

Ellis, P.R. and Hardman, J.A. (1987) Non-chemical contributors to an integrated programme for the control of carrot fly. *IOBC/WPRS Proceedings of the Working Group on Integrated Control in Field Vegetables*, Tune, Denmark, 21–23 September, 1987.

Ellis, P.R. and Hardman, J.A. (1992) Pests of umbelliferous crops. In: McKinlay, R.G. (ed.) *Vegetable Crop Pests.* CRC Press, Boca Raton, Florida, pp. 327–378.

Ellis, P.R., Freeman, G.H. and Hardman, J.A. (1984) Differences in the relative resistance of two carrot cultivars to carrot fly attack over five seasons. *Annals of Applied Biology* 105, 557–564.

Ellis, P.R., Hardman, J.A., Cole, R.A. and Phelps, K. (1987) The complementary effect of plant resistance and the choice of sowing and harvest times in reducing carrot fly (*Psila rosae*) damage to carrots. *Annals of Applied Biolology* 111, 415–425.

Ellis, P.R., Saw, P.L. and Crowther, T.C. (1991) Development of carrot inbreds with resistance to carrot fly using a single seed descent programme. *Annals of Applied Biology* 119, 349–357.

Erickson, E.H., Peterson, C.E. and Werner, P. (1979) Honeybee foraging and resultant seed set among male-fertile and cytoplasmic male-sterile carrot inbreds and hybrid seed parents. *Journal of the American Society for Horticultural Science* 104, 935–638.

Esau, K. (1940) Developmental anatomy of the fleshy storage organ of *Daucus carota. Hilgardia* 13(5),175–226.

Espinoza, L., Sanchez, C.A. and Schueneman, T.J. (1993) Celery yield responds to phosphorus rate but not phosphorus placement on histosols. *HortScience* 28, 1168–1170.

Evers, A.-M. (1989a) Effects of different fertilization practices on the growth, yield and dry matter content of carrot. *Journal of Agricultural Science of Finland* 60, 135–152.

Evers, A.-M. (1989b) Effects of different fertilization practices on the carotene content of carrot. *Journal of Agricultural Science of Finland* 61, 7–14.

Falk, B.W. (1994) The etiology, epidemiology and control of carrot motley dwarf in California. In: *California Fresh Carrot Advisory Board 1993 Annual Report*. Dinuba, California, pp. 105–109.

Finch-Savage, W.E. (1984) The establishment of direct-sown germinating celery seeds in peat blocks. *Journal of Horticultural Science* 59, 87–93.

Finch-Savage, W.E. (1986) The effects of fluid drilling and seed covering on early carrot production under polyethylene mulch. *Annals of Applied Biology* 108, 431–439.

Finch-Savage, W.E. and Cox, C.J. (1982a) Effects of adding plant nutrients to the gel carrier used for fluid drilling early carrots. *Journal of Agricultural Science (Cambridge)* 99, 295–303.

Finch-Savage, W.E. and Cox, C.J. (1982b) A cold treatment technique to improve the germination of vegetable seed prior to fluid drilling. *Scientia Horticulturae* 16, 301311.

Finch-Savage, W.E. and Pill, W.G. (1990) Improvement of carrot crop establishment by combining seed treatment with increased seedbed moisture availability. *Journal of Agricultural Science* 115, 75–81.

Francois, L.E. and West, D.W. (1982) Reduction in yield and market quality of celery caused by soil salinity. *Journal of the American Society for Horticultural Science* 107, 952–954.

Freeman, R.E. and Simon, P.W. (1983) Evidence for simple genetic control of sugar type in carrot (*Daucus carota* L.). *Journal of the American Society for Horticultural Science* 108, 50–54.

Frese, L. (1983) Resistance of the wild carrot *Daucus carota* ssp. *hispanicus* to the root-knot nematode *Meloidogyne hapla*. *Journal of Plant Diseases Protection* 81, 396–403.

Frese, L. and Webser, W.E. (1984) A method for the differentiation of breeding material of carrot with partial resistance to *Meloidogyne hapla*. *Journal of Plant Diseases Protection* 91, 396–403.

Frison, E.A. and Serwinski, J. (1995) *Directory of European Institutions Holding Crop Genetic Resources Collections*. IPGRI, Rome.

Gilbertson, R.L. (1996) Continued studies of carrot black rot and new methods for identification of *Alternaria* spp. In: *California Fresh Carrot Advisory Board 1995 Annual Report*. Dinuba, California, pp. 71–79.

Gray, D. (1984) The performance of carrot seeds in relation to their viability. *Annals of Applied Biology* 104, 559–565.

Gray, D. and Steckel, J.R.A. (1980) Studies on the sources of variation in plant weight in *Daucus carota* (carrot) and the implications for seed production techniques. In: Hebblewaite, P.H. (ed.) *Seed Production*. Butterworths, London, pp. 475–484.

Gray, D. and Steckel, J.R.A. (1983a) Some effects of umbel order and harvest date on carrot seed variability and seedling performance. *Journal of Horticultural Science* 58, 73–82.

Gray, D. and Steckel, J.R.A. (1983b) Seed quality in carrots: the effects of seed crop plant density, harvest date and seed grading on seed and seedling variability.

Journal of Horticultural Science 58, 393–401.

Gray, D. and Steckel, J.R.A. (1983c) Some effect of ripening and drying of carrot seed (*Daucus carota*) crops on germination. *Annals of Applied Biology* 101, 397–406.

Gray, D. and Ward, J.A. (1985) Relationship between seed weight and endosperm characteristics in carrot. *Annals of Applied Biology* 106, 379–384.

Gray, D., Steckel, J.R.A. and Ward, J.A. (1983) Studies on carrot seed production: effects of plant density on yield and components of yield. *Journal of Horticultural Science* 58, 83–90.

Gray, D., Ward, J.A. and Steckel, J.R.A. (1984) Endosperm and embryo development in *Daucus carota* L. *Journal of Experimental Botany* 35, 459–465.

Gray, D., Steckel, J.R.A., Jones, S.R. and Senior, D. (1986) Correlations between variability in carrot (*Daucus carota* L.) plant weight and variability in embryo length. *Journal of Horticultural Science* 61(1), 71–80.

Guzman, V.L. (1969) Celery yield and quality in relation to methods of fertilizer application and climate. *Soil and Crop Science Society of Florida* 29, 301–305.

Guzman, V.L., Burdine, H.W., Harris Jr., E.D., Orsenigo, J.R., Showalter, R.K., Thayer, P.L., Winchester, J.A., Wolf, E.A., Berger, R.D., Genung, W.G. and Zitter, T.Z. (1973) Celery production on organic soils of south Florida. *Florida Agricultural Experimental Station Bulletin* 757.

Hamilton, H.A. and Bernier, R. (1975) N-P-K Fertilizer effects on yield, composition and residues of lettuce, celery, carrot and onion grown on an organic soil in Quebec. *Canadian Journal of Plant Science* 55, 453–461.

Havis, L. (1939) Anatomy of the hypocotyl and roots of *Daucus carota*. *Journal of Agricultural Research* 58, 557–564.

Hayward, H.E. (1938) *The Structure of Economic Plants*. Macmillan Publishers, New York.

Heath-Pagliuso, S., Pullman, G. and Rappaport, L. (1988) Somaclonal variation in celery: screening for resistance to *Fusarium oxysporum* f.sp. *apii*. *Theoretical and Applied Genetics* 75, 446–451.

Heatherbell, D.A., Wrolstad, R.E. and Libbey, L.M. (1971) Carrot volatiles. 1. Characterization and effects of canning and freeze drying. *Journal of Food Science* 36, 219–224.

Hermann, M. (1997) Arracacha. In: Hermann, M. and Heller, J. (eds) *Andean Roots and Tubers: Ahipa, Arracacha, Maca and Yacon. Promoting the Conservation and Use of Underutilized and Neglected Crops*. Institute of Plant Genetics and Crop Plant Research, Gatersleben/International Plant Genetic Resources Institute, Rome.

Heywood, V.H. (1971) Chemosystematic studies in *Daucus* and allied genera. *Boissiera* 19, 289–295.

Heywood, V.H. (1978) Multivariate taxonomic synthesis of the tribe Caucalideae. *Proceedings of Symposium International sur les Umbelliferes*, pp. 727–736.

Heywood, V.H. (1983) Relationship and evolution in the *Daucus carota* complex. *Israel Journal of Botany* 32, 51–65.

Hodge, W.H. (1960) Yareta, fuel umbellifer of the Andean Puna. *Economic Botany* 14, 112–118.

Hole, C.C. (1996) Carrots. In: Zamski, E. and Schaffer, A.A. (eds) *Photoassimilate Distribution in Plants and Crops – Source–Sink Relationships*. Marcel Dekker, New York, pp. 671–690.

Hole, C.C. and Dearman, J. (1991) Carbon economy of carrots during initiation of the storage root in cultivars contrasting in shoot:root ratio at maturity. *Annals of Botany* 68, 427–434.

Hole, C.C. and Dearman, J. (1993) The effect of photon flux density on distribution of assimilate between shoot and storage root of carrot, red beet and radish. *Scientia Horticulturae* 55, 213–225.

Hole, C.C. and Sutherland, R.A. (1990) The effect of photon flux density and duration of the photosynthetic period on growth and dry matter distribution in carrot. *Annals of Botany* 65, 63–69.

Hole, C.C., Barnes, A., Thomas, T.H., Scott, P.A. and Rankin, W.E.F. (1983) Dry matter distribution between the shoot and storage root of carrot (*Daucus carota* L.). I. Comparison of varieties. *Annals of Botany* 51, 175–187.

Hole, C.C., Morris, G.E.L. and Cowper, A.S. (1987) Distribution of dry matter betweeen shoot and storage root of field grown carrot cultivars. II. Relationship between initiation of leaves and storage roots in different cultivars. *Journal of Horticultural Science* 62, 343–349.

Holland, B., Welch, A.A., Unwin, I.D., Buss, D.H., Paul, A.A. and Southgate, D.A.T. (1992) *McCance and Widdowson's The Composition of Foods*, 5th edn. Royal Society of Chemistry, Cambridge.

Honma, S. (1959a) A method for celery hybridization. *Proceedings of the American Society for Horticultural Science* 73, 345–348.

Honma, S. (1959b) A method for evaluating resistance to bolting in celery. *Proceedings of the American Society for Horticultural Science* 74, 506–513.

Honma, S. and Lacy, M. (1980) Hybridization between pascal celery and parsley. *Euphytica* 29, 801–805.

Huang, S.P., Della Vecchia, P.T. and Ferreira, P.E. (1986) Varietal response and estimates of heritability of resistance to *Meloidogyne javanica* in carrots. *Journal of Nematology* 18, 496–501.

Huestis, G. (1992) Genetic mapping in celery (*Apium graveolens* L.) and its wild allies based on restriction fragment length polymorphisms. PhD thesis, University of California, Davis, USA.

Huestis, G.M., McGrath and Quiros, C.F. (1993) Development of genetic markers in celery based on restriction fragment length polymorphisms. *Theoretical and Applied Genetics* 85, 889–896.

Jacobsohn, R. and Globerson, D. (1980) *Daucus carota* (carrot) seed quality. I. Effects of seed size on germination, emergence and plant growth under subtropical conditions. II. The importance of the primary umbel in carrot-seed production. In: Hebblethwaite, P.D. (ed.) *Seed Production*. Butterworths, London, pp. 637–646.

Janse, J.D. (1988) A *Streptomyces* species identified as the cause of carrot scab. *Netherlands Journal of Plant Pathology* 94, 303–306.

Johnson, A.E., Nursten, H.E. and Williams, A.A. (1971). Vegetable volatiles: a survey of components identified: Part II. *Chemistry and Industry* 1971, 1212–1224.

Kalloo, G. (1993a) Parsnip *Pastinaca sativa* L. In: Kalloo, G. and Bergh, B.O. (eds) *Improvement of Vegetable Crops*. Pergamon Press, Oxford, pp. 485–486.

Kalloo, G. (1993b) Turnip-rooted parsley *Petroselium crispum* Mill. Nyn., *P. crispum* var. *tuberosum*. In: Kalloo, G. and Bergh, B.O. (eds) *Improvement of Vegetable Crops*. Pergamon Press, Oxford, pp. 573–577.

Kemp, W.G. and Barr, D.J.S. (1978) Natural occurrence of tobacco necrosis virus in a rusty root disease complex of *Daucus carota* in Ontario. *Phytopathologische*

Zeitschrift 91, 203–217.

Lafuente, M.T., Cantwell, M., Rubatzky, V. and Yang, S.F. (1991) Factors influencing isocoumarin formation in carrots exposed to ethylene. In: A. Bonnet (ed.) *Eucarpia Carrot 91*, Avignon-Montfavet (France) 18–20 June, 1991, pp. 117–125.

Lafuente, M.T., Lopez-Galvez, G., Cantwell, M. and Yang, S.F. (1996) Factors influencing ethylene induced isocoumarin formation and increased respiration in carrots. *Journal of the American Society for Horticultural Science* 121, 537–542.

Lamprecht, H. (1961) Ein bastard, *Apium graveolens* L. × *Petroselinum hortense* Hoffm. *Agri. Hortique Genetica* 19, 223–224.

Lazcano, C.A., Dainello, F.J., Pike, L.M., Miller, M.E., Brandenberger, L. and Baker, L.R. (1998) Seed lines, population density and root size at harvest affect quality and yield of cut-and-peel baby carrots. *HortScience* 33, 972–975.

Lebeda, A. and Coufal, J. (1987) Evaluation of susceptibility of *Daucus carota* varieties to natural infection with *Erysiphe heraclei*. *Arch. Züchtungsforsch* 17, 73–76.

Lebeda, A., Coufal, J. and Kvasnicka, P. (1988) Evaluation of field resistance of *Daucus carota* cultivars to *Cercospora carotae* (carrot leaf spot). *Euphytica* 39, 285–288.

Lee, C.Y. (1986) Changes in carotenoid content of carrots during growth and post-harvest storage. *Food Chemistry* 20, 285–293.

Lester, O.E., Baker, L.R. and Kelly, J.F. (1982) Physiology of sugar accumulation in carrot breeding lines and cultivars. *Journal of the American Society for Horticultural Science* 107, 381–387.

Lewak, S. and Rudnicki, R.M. (1977) After ripening in cold-requiring seeds. In: Khan, A.A. (ed.) *The Physiology and Biochemistry of Seed Dormancy and Germination*. Elsevier/North Holland, New York, pp. 193–208.

Lewis, B.G., Davies, W.P. and Garrod, B. (1981) Wound healing in carrot roots in relation to infection by *Mycocentrospora acerina*. *Annals of Applied Biology* 99, 35–42.

Lockheart, C.L. and Delbridge, R.W. (1974) Control of storage diseases of carrots with postharvest fungicide treatments. *Canadian Plant Disease Survey* 54, 52–54.

Mackevic, V.I. (1929) The carrot of Afghanistan. *Bulletin of Applied Botany, Genetics and Plant Breeding* 20, 517–562.

Madjarova, D.J. and Bubarova, M.G. (1978) New forms obtained by hybridization of *Apium graveolens* and *Petroselinum hortense*. *Acta Horticulturae* 73, 65–72.

Mathias, M.E. (1971) Systematic survey of New World Umbelliferae. In: Heywood, V.H. (ed.) *The Biology and Chemistry of the Umbelliferae*. Academic Press, New York, pp. 13–29.

Maynard, D.N. and Hochmuth, G.J. (1997) *Knott's Handbook for Vegetable Growers*, 4th edn. John Wiley & Sons, New York.

Mazza, G. and Miniati, E. (1993) Roots, tubers and bulbs. In: *Anthrocyanins in Fruits, Vegetables, and Grains*. CRC Press, Boca Raton, Florida, p. 265.

McGarry, A. (1993) Influence of water status on carrot (*Daucus carota* L.) fracture properties. *Journal of Horticultural Science* 68, 431–437.

McKee, J.M.T. and Morris, G.E.L. (1986) Growth regulator effects on storage root development in carrot. *Plant Growth Regulators* 2, 359–369.

McLellan, M.E., Cash, J.N. and Gray, J.I. (1983) Characterization of the aroma of raw carrots (*Daucus carota* L.) with the use of factor analysis. *Journal of Food Science* 48, 71–72.

McNeal, B.L. and Pratt, P.F. (1978) Leaching of nitrate from soils. In: Pratt, P.F. (ed.) *Management of Nitrogen in Irrigated Crops.* University of California, Riverside, pp. 195–230.

McPharlin, I.R., Jeffer, R.C. and Weissberg, R. (1994) Determination of the residual value of phosphate and soil test phosphorus calibration for carrots on a Karrakatta sand. *Communications in Soil Science and Plant Analysis* 25, 489–500.

Michalik, B, Simon, P.W. and Gabelman, W.H. (1992) Assessing susceptibility of carrot roots to bacterial soft rot. *HortScience* 27, 1020–1022.

Munger, H.M. (1987) Adaptation and breeding of vegetable crops for improved human nutrition. In: Quebedeaux, B. and Bliss, F.A. (eds) *Horticulture and Human Health.* Prentice Hall, Englewood Cliffs, New Jersey, pp. 177–184.

Munger, H.M. and Newhall, A.G. (1952) Emerson pascal, a blight resistant celery. *New York Agricultural Experimental Station Farm Research* 17, 11.

Murata, M. and Orton, T. (1984) G-banding like differentiation in meiotic prometaphase chromosomes of celery. *Journal of Heredity* 75, 225–228.

Ochoa, O. and Quiros, C.F. (1989) *Apium* wild species: Novel sources for resistance to late blight in celery. *Plant Breeding* 102, 317–321.

Ochoa, O., D'Antonio, V. and Quiros, C.F. (1986) Techniques for water emasculation and cut seedstalk pollination in celery. *HortScience* 21, 1455–1456.

O'Hare, S.K., Locascio, S., Forbes, R., White, J.M., Hensel, D., Shumaker, J. and Dangler, J.M. (1983) Root crops and their biomass potential in Florida. *Proceedings of the Soil and Crop Science Society of Florida* 42, 13–17.

Olymbios, C.M. (1973) Physiological studies on the growth and development of carrot, *Daucus carota* L. PhD thesis, University of London, UK.

Orton, T.J., Durgan, M.E. and Hulbert, S.H. (1984a) Studies on the inheritance of resistance to *Fusarium oxysporum* f.sp. *apii* in celery. *Plant Disease* 68, 574–578.

Orton, T.J., Hulbert, S.H., Durgan, M.E. and Quiros, C.F. (1984b) UC1, Fusarium yellows-resistant celery breeding line. *HortScience* 19, 594.

Otsuka, H. and Take, T. (1969) Sapid components in carrot. *Journal of Food Science* 34, 392–394.

Page, E.R. and Gerwitz, A. (1969) Phosphate uptake by lettuces and carrots from different soil depths in the field. *Journal of Science and Food Agriculture* 20, 85–90.

Pan, H.P. (1961) Bitterness in celery. *Journal of Food Science* 26, 337–344.

Parsons, C.S. (1960) Effects of temperature, packaging and sprinkling on the quality of stored celery. *Proceedings of the American Society for Horticultural Science* 75, 463–469.

Peacock, L. (1991) Effect on weed growth of short-term cover over organically grown carrots. *Biological Agriculture and Horticulture* 7, 271–279.

Pepkowitz, L.P., Larson, R.E., Gardner, J. and Owens, G. (1944) The carotene and ascorbic acid concentration in vegetable varieties. *Plant Physiology* 19, 615–626.

Peterson, C.E. and Simon, P.W. (1986) Carrot breeding. In: Bassett, M.J. (ed.) *Breeding Vegetable Crops.* AVI, Westport, Connecticut, pp. 321–356.

Pfleger, F.L., Harman, G.E. and Marx, G.A. (1974) Bacterial blight of carrots: interaction of temperature, light and inoculation procedures on disease development of various carrot cultivars. *Phytopathology* 64, 746–749.

Phan, C.T. and Hsu, H. (1973) Physical and chemical changes occurring in the carrot root during growth. *Canadian Journal of Plant Science* 53, 629–634.

Pharr, D.M., Stoop, J.M.H., Williamson, J.D., Studer Feusi, M.E., Massel, M.O. and Conkling, M.A. (1995) The dual role of mannitol as osmoprotectant and photoassimilate in celery. *HortScience* 30, 1182–1188.

Pieczarka, D.J. (1981) Shallow planting and fungicide application to control Rhizoctonia stalk rot of celery. *Plant Disease* 65, 879–880.

Pink, D.A., Walkey, D.G., Stanley, A.R., Carter, P.J., Smith, B.M., Mee, C. and Bolland, C.J. (1983) Genetics of disease resistance. *National Vegetable Research Station Annual Report*, Wellesbourne, Warwick, UK, pp. 58–59.

Prabhaker, M., Srinivas, K. and Hegde, D.M. (1991) Effect of irrigation regimes and nitrogen fertilization on growth, yield, nitrogen uptake and water use of carrot (*Daucus carota* L.). *Gartenbauwissenschaft* 56(5), 206–209.

Pressman, E. and Negbi, M. (1980) The effect of day length on the response of celery to vernalization. *Journal of Experimental Botany* 31, 291–1296.

Pressman, E. and Negbi, M. (1987) Interaction of day length and applied gibberellins on stem growth and leaf production in three varieties of celery. *Journal of Experimental Botany* 38, 968–971.

Pressman, E., Negbi, M., Sachs, M. and Jacobsen, J.V. (1977) Varietal differences in light requirements for germination of celery (*Apium graveolens* L.) seeds and the effects of thermal and solute stress. *Australian Journal of Plant Physiology* 4, 821–831.

Pryor, B.M., Davis, R.M. and Gilbertson, R.L. (1994) Detection and eradication of *Alternaria radicina* on carrot seed. *Plant Disease* 78, 452–456.

Pryor, B.M., Davis, R.M. and Gilbertson, R.L. (1998) Detection of soilborne *Alternaria radicina* and its occurrence in California carrot fields. *Plant Disease* 82, 891–895.

Punja, Z.K. (1987) Mycelial growth and pathogenesis of *Rhizoctonia carotae* on carrot. *Canadian Journal of Plant Pathology* 9, 24–31.

Punja, Z.K., Carter, J.D., Campbell, G.M. and Rossell, E.L. (1986) Effects of calcium and nitrogen fertilizers, fungicides and tillage practices on incidence of *Sclerotium rolfsii* on processing carrots. *Plant Disease* 70, 819–824.

Quiros, C.F. (1987) Breeding celery for disease resistance and improved quality. *1986–87 California Celery Research Advisory Board Annual Report*. Dinuba, California.

Quiros, C.F. (1990) Breeding celery for disease resistance. *1989–90 California Celery Research Advisory Board Annual Report*. Dinuba, California.

Quiros, C.F. (1993) Celery – *Apium gravelons* L. In: Kalloo, G. and Bergh, B.O. (eds) *Genetic Improvement of Vegetable Crops*. Pergamon Press, Oxford, pp. 523–534.

Quiros, C.F. (1997) Celery genetics and breeding for disease resistance. In: *Annual Report 1996–97 California Celery Research Advisory Board*. Dinuba, California, pp. 7–14.

Quiros, C.F., Rugama, A., Dong, Y.Y. and Orton, T.J. (1986) Cytological and genetic studies of a male sterile celery. *Euphytica* 35, 867–875.

Quiros, C.F., Douches, D. and D'Antonio, V. (1987) Inheritance of annual habit in celery: cosegregation with isozyme and anthocyanin markers. *Theoretical and Applied Genetics* 74, 203–208.

Ramin, A.A. and Atherton, J.G. (1991a) Manipulation of bolting and flowering in celery (*Apium graveolens* L. var. *dulce*). I. Effects of chilling during germination and seed development. *Journal of Horticultural Science* 66, 435–441.

Ramin, A.A. and Atherton, J.G. (1991b) Manipulation of bolting and flowering in celery (*Apium graveolens* L. var. *dulce*). II. Juvenility. *Journal of Horticultural Science* 66, 709–717.

Ramin, A.A. and Atherton, J.G. (1994) Manipulation of bolting and flowering in celery *Apium graveolens* L. var. *dulce*). III. Effects of photoperiod and irradiance. *Journal of Horticultural Science* 69(5), 861–868.

Reyes, A.A. (1988) Suppression of *Sclerotina sclerotiorum* and watery soft rot of celery by controlled atmosphere storage. *Plant Disease* 72, 790–792.

Roberts, P.A. (1998) Identification of gene sources for resistance to root knot nematodes attacking carrots in California. In: *1997 Annual Report California Fresh Carrot Research Advisory Board.* Dinuba, California, pp. 61–63.

Robinson, F.E. (1969) Carrot population density and yield in a arid environment. *Agronomy Journal* 61, 499–500.

Robinson, R.W. (1954) Seed germination problems in the Umbelliferae. *Botany Review* 20, 531–550.

Roelofse, E.W., Hand, D.W. and Hall, R.L. (1990) The effects of temperature and 'night-break' lighting on the development of glasshouse celery. *Journal of Horticultural Science* 65, 297–307.

Runham, S.R., Davies, J.S. and Rickard, P.C. (1992) *In-situ* field storage of carrots. *Journal of the Royal Agricultural Society of England* 153, 145–151.

Ryall, A.L. and Lipton, W.J. (1972) *Handling, Transportation, and Storage of Fruits and Vegetables*, Vol. I, *Vegetables and Melons*. AVI, Westport, Connecticut.

Ryan, E.W. (1973) The effects of cultivar, time of sowing and inter-row spacing on canker and yield of parsnips. *Acta Horticulturae* 27, 144–149

Sachs, M. and Ryski, I. (1980) The effects of temperature and daylength during the seedling stage on flower-stalk formation in field-grown celery. *Scientia Horticulturae* 12, 231–242.

Sáenz Laín, C. (1981) Research on *Daucus* L. (Umbelliferae). *Anales sel Instituto Botanico A.J. Cavanilles* 37, 481–534.

Salunkhe, D.K. (1961) Gamma irradiation effects on fruits and vegetables. *Economic Botany* 15, 28– 56.

Salter, P.J., Currah, I.E. and Fellows, J.R. (1981) Studies on some sources of variation in carrot root weight. *Journal of Agricultural Science* 96, 549–556.

Sanchez, C.A., Burdine, H.W. and Guzman, V.L. (1990) Soil testing and plant analysis as guides for the fertilization of celery on histosols. *Soil and Crop Science Fla. Proceedings* 49, 69–72.

Schultz, B., Westphal, L. and Wricke, G. (1994) Linkage groups of isozyme, RFLP and RAPD markers in carrot (*Daucus carota* L. *sativus*). *Euphytica* 74, 67–76.

Senalik, D. and Simon, P.W. (1987) Quantifying intra-plant variation of volatile terpenoids in carrot. *Phytochemistry* 26, 1975–1979.

Shadbolt, C.A. and Holm, L.G. (1956) Some quantitative aspects of weed competition in vegetable crops. *Weeds* 4, 111–123.

Sharma, A.K. and Bhattacharyya, N.K. (1954) Further investigations on several genera of Umbelliferae and their relationship. *Genetica* 30, 1–68.

Simon, P.W. (1984) Genetic effects on the flavor of stored carrots. *Acta Horticulturae* 163, 137–142.

Simon, P.W. (1987) Genetic improvement of carrots for meeting human nutritional needs. In: Quebedeaux, B. and Bliss, F.A. (eds) *Horticulture and Human Health.* Prentice Hall, Englewood Cliffs, New Jersey, pp. 208–214.

Simon, P.W. (1990) Carrots and other horticultural crops as a source of provitamin A carotenes. *HortScience* 25, 1495–1499.

Simon, P.W. (1992) Genetic improvement of vegetable carotene content. In: *Biotechnology and Nutrition – Proceedings of the Third International Symposium.* Butterworth-Heinemann, London, pp. 291–300.

Simon, P.W. (1996) Inheritance and expression of purple and yellow storage root color in carrot. *Journal of Heredity* 87(1), 63–66.

Simon, P.W. and Lindsay, R.C. (1983) Effects of processing on objective and sensory variables of carrots. *Journal of the American Society for Horticultural Science* 108, 928–931.

Simon, P.W. and Strandberg, J.O. (1998) Diallel analysis of resistance in carrot to *Alternaria* leaf blight. *Journal of the American Society for Horticultural Science* 123, 412–415.

Simon, P.W. and Wolff, X.Y. (1987) Carotene in typical and dark orange carrots. *Journal of Agricultural Food Chemistry* 35, 1017–1022.

Simon, P.W., Peterson, C.E. and Lindsay, R.C. (1980a) Correlations between sensory and objective parameters of carrot flavor. *Journal of Agricultural Food Chemistry* 28, 559–562.

Simon, P.W., Peterson, C.E. and Lindsay, R.C. (1980b) Genetic and environmental influences on carrot flavor. *Journal of the American Society for Horticultural Science* 105, 416–420.

Simon, P.W., Peterson, C.E. and Lindsay, R.C. (1982) Genotype, soil and climate effects on sensory and objective components of carrot flavor. *Journal of the American Society for Horticultural Science* 107, 644–648.

Simon, P.W., Wolff, X.Y., Peterson, C.E., Kammerlohr, D.S., Rubatzky, V.E., Strandberg, J.O., Bassett, M.J. and White, J.M. (1989) High carotene mass carrot population. *HortScience* 24, 174.

Simon, P.W., Peterson, C.E. and Gabelman, W.H. (1990) B493 and B9304. Carrot inbreds for use in breeding, genetics and tissue culture. *HortScience* 25, 815.

Small, E. (1978) A numerical taxonomic analysis of the *Daucus carota* complex. *Canadian Journal of Botany* 56, 248–276.

Smith, P.R. (1966) Seed transmission of *I. pastinacae* in parsnip and its elimination by a steam air treatment. *Australian Journal of Experimental Agriculture* 6, 441–444.

Smith, R.B. and Reyes, A.A. (1988) Controlled atmosphere storage of Ontario-grown celery. *Journal of the American Society for Horticultural Science* 113, 390–394.

Sondheimer, E., (1957) The isolation and identification of 3-methyl-6-methoxy-8--hydroxy-3, 4-diyhydroscoumarin from carrots. *Journal of the American Chemical Society* 79, 5063–5066.

St. Pierre, M.D. and Bayer, R.J. (1991) The impact of domestication on the genetic variability in the orange carrot, cultivated *Daucus carota* ssp. *sativus* and the genetic homogeneity of various cultivars. *Theoretical and Applied Genetics* 82, 249–253.

Stark, J.C., Jarrell, W.M. and Letey, J. (1982) Relationship between growth and nitrogen fertilization of celery. *HortScience* 17, 754–755.

Stark, J.C., Jarrell, W.M. and Letey, J. (1983) Evaluation of irrigation-nitrogen management practices for celery using continuous-variable irrigation. *Soil Science Society of America Journal* 47, 95–98.

Steingrover, E. (1981) The relationship between cyanide-resistant root respiration and the storage of sugars in the taproot of *Daucus carota* L. *Journal of Experimental Botany* 32, 911–919.

Stommel, J.R. and Simon, P.W. (1989) Phenotypic recurrent selection and heritability estimates for total dissolved solids and sugar type in carrot. *Journal of the American Society for Horticultural Science* 114, 695–699.

Strandberg, J.O. (1983) Infection and colonization of inflorescences and mericarps of carrot by *Alternaria dauci*. *Plant Disease* 67, 1351–1353.

Strandberg, J.O. and White, J.M. (1978) *Cercospora apii* damage of celery – effect of plant spacing and growth on raised beds. *Phytopathology* 68, 223–226.

Sutton, J.C. (1975) *Pythium* spp. produce rusty root of carrot in Ontario. *Canadian Journal of Plant Science* 55, 139–143.

Tanne, I. and Cantliffe, D.J. (1989) Seed treatments to improve rate and uniformity of celery seed germination. *Proceedings of the Florida State Horticultural Society* 102, 319–322.

Thomas, T.H. (1989) Gibberellin involvement in dormancy-break and germination of seeds of celery (*Apium graveolens* L.). *Plant Growth Regulation* 8, 255–261.

Thomas, T.H. (1994) Responses of Florence fennel (*Foeniculum vulgare azoricum*) seeds to light, temperature and gibberellin $A_{4/7}$. *Plant Growth Regulation* 14, 139–143.

Thomas, T.H. and O'Toole, D.F. (1980) The effects of environment and chemical treatments on the production of some vegetable seeds. In: Hebblethwaite, P.D. (ed.) *Seed Production*. Butterworths, London, pp. 501–512.

Thomas, T.H., Palevitch, D., Biddington, N.L. and Austin, R.B. (1975) Growth regulators and the phytochrome-mediated dormancy of celery seeds. *Physiologia Plantarum* 35, 101–106.

Thomas, T.H., Biddington, N.L. and Palevitch, D. (1978a) Improving the performance of pelleted celery seeds with growth regulator treatments. *Acta Horiculturae* 83, 235–243.

Thomas, T.H., Gray, D. and Biddington, N.L. (1978b) The influence of the position of the seed on the mother plant on seed and seedling performance. *Acta Horticulturae* 83, 57–66.

Thomas, T.H., Biddington, N.L. and Palevitch, D. (1978c) The role of cytokinins in the phytochrome mediated germination of dormant inbibed celery (*Apium graveolens*) seeds. *Photochemistry and Photobiology* 27, 231–236.

Thomas, T.H., Biddington, N.L. and O'Toole, D.F. (1979) Relationship between the position on the parent plant and dormancy characteristics of seed of three cultivars of celery. *Physiologia Plantarum* 45, 492–496.

Toll, J. and van Sloten, D.H. (1982) *Directory of Germplasm Collections 4, Vegetables*. IBPGR Secretariat, FAO, Rome.

Townsend, G.R., Emerson, R.A. and Newhall, A.G. (1946) Resistance to *Cercospora apii* Fres., in celery (*Apium graveolens* var. *dulce*). *Phytopathology* 36, 980–982.

Trumble, J.T. and Quiros, C.F. (1988) Antixenotic and antibiotic resistance in *Apium* species to *Lyriomiza trifolii*. *Journal of Economic Entomology* 81, 602–607.

Van Den Berg, L. (1981) The role of humidity, temperature and atmospheric composition in maintaining vegetable quality during storage. *American Chemical Society Symposium Series*170, 96–107.

Van Den Berg, L. and Lentz, C.P. (1966) Effect of temperature, relative humidity and atmospheric composition on changes in quality of carrots during storage. *Food Technology* 20, July, 104–107 (954–957).

Vavilov, N.I. (1951) The origin, variation, immunity and breeding of cultivated plants. *Chron. Botany* 13, 1–366. [Translated from Russian by Starr Chester.]

Venter, F. (1979) Nitrate content of carrots as influenced by fertilization. *Acta Horticulturae* 93, 163–172.

Vilmorin, R.L. (1950) Pascal celery and its origin. *Journal of the New York Botanical Garden* 51, 39–41.

Vivek, B.S. (1997) Carrot (*Daucus carota* L.) molecular markers: investigations into *Daucus* phylogeny, chloroplast DNA inheritance and genetic mapping. PhD. thesis, University of Wisconsin, Madison.

Vivek, B.S. and Simon, P.W. (1998a) Genetic relationships and diversity in carrot and other *Daucus* taxa based on nuclear restriction fragment length polymorphism (nRFLPs). *Journal of the American Society for Horticultural Science* 123, 1053–1057.

Vivek, B.S. and Simon, P.W. (1998b) Phylogeny and relationships in *Daucus* based on restriction fragment length polymorphisms (RFLPs) of the chloroplast and mitochondrial genomes. *Euphytica* 105, 183–189.

Vivek, B.S. and Simon, P.W. (1999) Linkage relationships among molecular markers and storage root traits of carrot (*Daucus carota* L. *sativus*). *Theoretical and Applied Genetics* (in press).

Vivek, B.S., Ngo, Q.A. and Simon, P.W. (1999) Evidence for maternal inheritance of the chloroplast genome in cultivated carrot (*Daucus carota* L. *sativus*). *Theoretical and Applied Genetics* 98, 669–672.

Vrain, T.C. and Baker, L.R. (1980) Reaction of hybrid carrot cultivars to *Meloidogyne hapla*. *Canadian Journal of Plant Pathology* 2, 163–168.

Walkey, D.G.A. and Cooper, V.C. (1971) Effect of western celery mosaic on celery crops in Britain and occurrence of the virus in Umbelliferous weeds. *Plant Disease Reporter*. 55, 268–271.

Wareing, P.F. and Partick, J. (1975) Source–sink relations and the partition of assimilates in the plant. In: Cooper, J.P. (ed.) *Photosynthesis and Productivity in Different Environments.* Cambridge University Press, London, pp. 481–499.

Warren-Wilson, J. (1972) Control of crop processes. In: Ress, A.R., Cockshull, K.E., Hand, D.W. and Hurd, R.G. (eds) *Crop Processes in Controlled Environments.* Academic Press, London, pp.7–30.

Watson, M.A. (1990) Carrot motley dwarf virus. *Plant Pathology* 9, 133–134.

Weichmann, J. (1977) CA storage of celeriac. *Acta Horticulturae* 62, 109–118.

Weichmann, J. and Ammerseder, E. (1974) Influence of CA storage conditions on carbohydrate changes in carrots. *Acta Horticulturae* 38, 339–344.

Welch, J.E. and Grimball, E.L. (1947) Male sterility in carrot. *Science* 106, 594.

Wheeler, T.R., Morison, J.I.L., Ellis, R.H. and Hadley, P. (1994) The effects of CO_2, temperature and their interaction on the growth and yield of carrot (*Daucus carota* L.). *Plant, Cell and Environment* 17, 1275–1284.

White, J.G. (1988) Studies on the biology and control of cavity spot of carrots. *Annals of Applied Biology* 113, 259–268.

White, J.M. (1978) Soil preparation effects on compaction, carrot yield and root characteristics in organic soil. *Journal of the American Society for Horticultural Science* 103, 433–435.

White, J.M. (1992) Carrot yield when grown under three soil water concentrations. *HortScience* 27, 105–106.

White, J.M. and Strandberg, J.O. (1978) Early root growth of carrots in organic soil. *Journal of the American Society for Horticultural Science* 103, 344–347.

White, J.M. and Strandberg, J.O. (1979) Physical factors affecting root growth: water saturation of soil. *Journal of the American Society for Horticultural Science* 104, 414–416.

Yang, X.F. and Quiros, C.F. (1993) Identification and classification of celery cultivars with RAPD markers. *Theoretical and Applied Genetics* 86, 205–212.

Yang, X.F. and Quiros, C.F. (1994) Construction of genetic linkage map in celery using DNA-based markers. *Genome* 38, 36–44.

Zeevaart, J.A.D. (1978) Phytohormones and flower formation. In: Letham, D.S., Goodwin, P.B. and Higgins, T.J.V. (eds) *Phytohormones and Related Compounds: a Comprehensive Treatise*, Vol. 2. Elsevier/North Holland Biomedical Press, Amsterdam, pp. 291–338.

Zink, F.W. (1963) Rate of growth and nutrient absorption by celery. *Proceedings of the American Society for Horticultural Science* 82, 351–357.

Zukowska, E., Czeladzka, B. and Zabaglo, A. (1997) Differences in macroelements and nitrate content in the roots of some carrot genotypes. *Journal of Applied Genetics* 38A, 153–159.

Index